Sex on Six Legs

BOOKS BY MARLENE ZUK

Sexual Selections:
What We Can and Can't Learn
about Sex from Animals

Riddled with Life:
Friendly Worms, Ladybug Sex,
and the Parasites That Make Us
Who We Are

Sex on Six Legs:
Lessons on Life, Love,
and Language from the Insect World

Sex on Six Legs

LESSONS ON LIFE, LOVE, AND LANGUAGE FROM THE INSECT WORLD

MARLENE ZUK

HOUGHTON MIFFLIN HARCOURT
Boston New York

www.hmhbooks.com

Library of Congress Cataloging-in-Publication Data
Zuk, M. (Marlene)
Sex on six legs / Marlene Zuk.
p. cm.
ISBN 978-0-15-101373-9
1. Insects — Behavior. 2. Insects — Sexual behavior. I. Title.
QL496.Z85 2011
595.715 — dc22 2010025829

Book design by Melissa Lotfy

Printed in the United States of America

DOC 10 9 8 7 6 5 4

Contents

Sex on Six Legs

INTRODUCTION

Life on Six Legs

> Two-legged creatures we are supposed to love as we love ourselves. The four-legged, also, can come to seem pretty important. But six legs are too many from the human standpoint.
>
> — JOSEPH W. KRUTCH

PEOPLE are more afraid of insects than they are of dying, at least if you believe a 1973 survey published in *The Book of Lists.* Only public speaking and heights exceeded the six-legged as sources of fear, although "financial problems" and "deep water" (presumably when one was immersed in it) tied with insects at number three. Dying came in at number six. I have no reason to expect that matters have changed much, and suspect that if spiders had been included with insects in the options, fear of the multilegged would have easily topped the chart. People have strong feelings about insects, and most of those feelings are negative.

And yet for centuries, some of the greatest minds in science have drawn inspiration from studying some of the smallest minds on earth. From Jean Henri Fabre to Charles Darwin to E. O. Wilson, naturalists have been fascinated by the lives of six-legged creatures that seem both frighteningly alien and uncannily familiar. Beetles and earwigs take care of their young, fireflies and crickets flash and chirp for mates, and ants construct elaborate societies, with internal politics that put the U.S. Congress to shame. And scientists — along with many backyard naturalists — keep on wanting to tell their stories.

It's not just that we publish scholarly journal articles about insects, or use them in our laboratories. Insects are special. Rats and mice are useful scientific tools, too, but although we personify them in fairy tales or cartoons, rodents are just not as compelling as bugs. Birds are beautiful, and we admire them and write poetry about their song and grace, but they don't get under our skin — literally or figuratively — the way that insects do. When it comes to insects, we write about *Life on a Little-Known Planet,* with *Bugs in the System.* We muse about *Little Creatures Who Run the World,* and we're only partly joking. Those of us who study insects are passionate about them in a way that can seem incomprehensible to outsiders. People get why Jane Goodall loves chimps; they are less sanguine about my fondness for earwigs.

Some of it, of course, is the sheer magnitude of almost everything about insects — they are more numerous than any other animal, making up over 80 percent of all species. Estimates of the number of kinds of insects vary wildly, because new ones are being discovered all the time, but there are at least a million, possibly as many as ten million, which means that you could have an "Insect of the Month" calendar and not need to re-use a species for well over eighty thousand years. Take that, pandas and kittens! At any one moment, say while you are reading this sentence, approximately ten quintillion (10,000,000,000,000,000,000) individual insects surround you in the world. All of that variety gives enormous scope for evolution to act upon. Think of all those species as possible ingredients for a menu in a vast natural restaurant. You can come up with a lot more living recipes with insects than with the paltry few thousand bird species out there. And then there is the sensationalism; nothing gets my students' attention like hearing about male honeybees' genitals exploding after sex, and everyone has shuddered over the female mantis eating her mate. Insects routinely do things that would put the most gruesome horror film to shame.

Of course, not everyone finds insects scary, *The Book of Lists* survey notwithstanding. Those books on insects find readers, the nature channels on TV often feature bugs, and in 2009 the London Zoo hosted a "Pestival," "celebrating insects in art, and the art of being an insect." It included art, lectures, discussions, and a celebration of all things entomological. It even featured a six-legged take on the recent death of pop star Michael Jackson: Japanese artist Noboru Tsubaki made a "Vegetable Wasp," described as "a kind of cocoon for Jackson to enable him to traverse between the world of the living and the dead." Whether this effort successfully put Jackson's spirit to rest or not, metamorphosis is a powerful, and not unwelcome, image for us noninsects to contemplate. When Isabella Rossellini made *Green Porno,* her series of short films on animal mating, she led off with insects: dragonfly, bee, mantis, housefly. They were compelling in a way that other animals are not.

So what is it that keeps us coming back to insects? Why do they inspire such strong emotions, and what can we learn about ourselves from watching their joint-legged lives? The newest discoveries in biology, about genomes and nerve cells and the evolutionary connections between them, are best revealed by insects. This book is my celebration of a world that is alien and familiar at the same time, an invitation to the latest news about insect lives. We are continuing to make extraordinary and important discoveries about insects, routinely even finding new species. I haven't seen *Green Porno,* but if the segment on dragonflies is up to date, it should include a shot of the male's jagged penis as it scoops out the sperm from a previous mate, replacing it with his own. Sperm competition, in which the sperm of multiple males battle inside a female's reproductive tract, was first discovered, and is best understood, in insects, and new aspects of it are being uncovered all the time.

Insects are even teaching us about mind control, and maybe even about consciousness itself. A tiny wasp called the emerald cockroach wasp can do what many renters cannot: direct the movements of a cockroach. The wasp does this not to rid a kitchen of scuttling invaders but to feed her brood. Many wasps provision their young by paralyzing other insects or spiders and carrying them back to the wasp's nest. The paralysis, as opposed to out and out killing of the prey, helps the prey stay fresh while the young wasp larva feasts on the flesh. Of course, paralyzed insects can't put themselves into the nest, so the wasp usually has to do all the heavy lifting, staggering under the weight of her groceries as she flies back to her young. Except, that is, in the case of the jewel wasp, so named for the glittery emerald sheen of her exoskeleton. The female wasp doesn't send the roach into an immobile stupor; instead, she makes it into a zombie via a judicious sting inside the roach's head, so that its nervous system, and legs, still function well enough to allow it to walk on its own. Then, as science writer Carl Zimmer describes, "The wasp takes hold of one of the roach's antennae and leads it, like a dog on a leash, to its doom."

For years scientists were mystified about the precision of this sinister manipulation of the nervous system. How could a single injection of venom manage to produce what neuroscientists Ram Gal and Frederic Libersat, from Ben-Gurion University of the Negev in Israel and the Université de la Méditerranée in France, called "a living yet docile" victim? Finally, in 2010, through a series of meticulous manipulations of the cockroach nervous system, including a kind of wasp-mimicking injection at various sites along the collections of nerve cells in the head, the researchers demonstrated that the drive to walk in response to most stimuli is seated in a tiny cluster of cells called the subesophageal ganglia. By poisoning just this minuscule part of the nervous system, the wasp is able, in Gal and Libersat's words, "to 'hijack the cockroach's free will.'" Zimmer refers to the discovery as finding "the seat of the cockroach soul." I am not so sure I buy the idea that roaches have souls to be found, nor that free will is residing in all those cockroaches lucky enough to miss an encounter with a jewel wasp, but then I am not sure about either of those things in humans, either. But the finding illustrates one of the most enthralling aspects of insects: they make difficult-to-grasp concepts, for example, souls and free will, satisfyingly literal. If we can get to a roach's motivation to walk by throwing a monkey wrench into a couple of cells, can the ability to find motivations for human behaviors be far behind?

Maybe you are convinced that insects are important simply because they invade our kitchens and crops, but you don't think they have any inherent magic. If you are one of those that think insects are important but not breathtaking, pests without inspiring passion, I want to change your mind. It's not just that insects are useful, even essential, given their role in pollination—providing what are now trendily called *ecosystem services*—or the use of their genetic information to cure malaria. Those practical reasons can make you need something, but not love it; no one denies our reliance on, say, soap, or drywall, but who wants to hear about their intricacies? Insects, on the other hand, can help us see another way of life, like a gloriously overblown version of cultural exchange. Travel is said to be broadening because it makes us realize that our way of doing things is not the only one, that people in other cultures live differently and get by just fine. Insects do that, too, only better. They too make us see that our way of life is not the only one—and I don't mean that we could be eating dung instead of cheeseburgers. I mean that it is possible to be unselfish without a moral code, sophisticated without an education, and beautiful wearing a skeleton on the outside. Insects can shake you in ways you never expected, and even more new discoveries about their lives have been made possible just in the last few years by the tools of genomics. So what do insects have that people haven't noticed?

Insects Are Equal Opportunity

INSECTS are the great equalizers. There is not a corner of the globe where people—rich, poor, old, or young—have not had some encounter with insects, even if only to swat a mosquito or crush a cockroach. Because of that ubiquity, insects are the easiest portal to the animal kingdom, an inadvertent reminder that other creatures live here besides us, whether we want them to or not. We are all in the same buzzing, crawling boat.

But this is not to bemoan that we are all dragged down by the assault of six-legged life on our crops or our persons, a kind of vermin-ridden misery loves company. Insects also provide a much more uplifting egalitarianism. If you want to learn about the natural world but are too young or too poor or otherwise lack an opportunity to study the stars or put droplets of pond scum under a microscope, bugs are always there for you. I grew up in the middle of Los Angeles in a modest neighborhood without creeks or woods or much in the way of encouragement to do a project for the science fair. But early on I discovered that if I lifted the hexagonal concrete pavers in the yard, ants would rush to and fro carrying their plump white pupae, and that the tiny spiky monsters on the rosebush would metamorphose into ladybugs. I reared the fritillary butterflies that lived on a passionflower vine in our yard, year after year, never tiring of watching as the eggs hatched into threadlike caterpillars that grew and grew inside my jars, eventually hanging upside down from a stick and becoming a gaudy spangled adult. No special equipment necessary, no need to venture anywhere my mother would disapprove of or that cost any money at all. And the results were just as compelling, maybe more so, than if I'd had a telescope or a dissecting kit or a way to watch the social lives of wolves.

This equal-opportunity entomology has been going on for centuries. Maria Sibylla Merian was a German-born painter whose work is rediscovered and shown every few decades; she was recently featured at the Getty Museum in Los Angeles. Merian documented, many years before the naturalists of the time, the life cycles of butterflies, moths, and other insects. Her work is exquisite from an aesthetic perspective, but what interests me more is that as a woman in the late seventeenth and early eighteenth centuries, she was able to make scientific contributions that would have been impossible in virtually any other field, simply by virtue of using the specimens from her own garden. She eventually traveled to Surinam to study the brilliantly colored insects of the steamy jungle, but that was after her interests had been firmly set. Although she, like many other women scientists and naturalists, faced opposition for her unfeminine activities, the accessibility of her subjects meant that she could keep doing the work she loved.

Interestingly, professional entomology has become one of the more male-dominated fields of biology, perhaps because of its connections with crop pest management and agribusiness, both of which tend to attract men. Regardless, it still appeals to children, both boys and girls, as my experience testifies. And even now it is not impossible to make important discoveries without a lot of technological gizmos. A group of scientists working in Brazil recently discovered that caterpillars parasitized by a wasp continue to make an unwitting sacrifice even after the wasp larvae have emerged from their host to pupate on a nearby stem. The ravaged caterpillar stands guard over the developing wasps and defends them against intruders with vigorous swings of its body, a most uncaterpillar-like behavior. Apparently the wasps exert a kind of mind control over their host that persists even after they leave it, doomed to die before it will ever become a moth.

This gruesome story has many arresting elements; most of the news coverage used words such as *voodoo* and *zombie,* as with the jewel wasp mentioned above. What I like most about it is that the scientists who discovered it were just watching the goings-on in a guava plantation, an illustration of what you can find if you are just paying attention. High tech has its place, of course, and I would hardly champion a return to simpler science or the eschewing of DNA sequencers. But I take great pleasure in the unifying ability of studying bugs. It's not just that insects level the playing field: they even supply the toys. The chapters that follow will let you play with them, will let you in on some remarkable new truths, in a way that would be impossible with most other fields of science.

Insects Are a Mirror

ALONG with all of their alien behavior, insects seem to do much of what people do: they meet, mate, fight, and part, and they do so with what looks like love or animosity. Dung beetles take care of their helpless squirming young, doing almost everything human mothers do, short of giving their baby a bottle — or parking it in front of the television. Ants keep aphid "cattle," moving their herd from place to place and milking the honeydew the aphids produce. Bees convey the location of food using symbols. Unlike any other nonhuman animal, some insects live in sophisticated hierarchical societies, with specialized tasks assigned to different individuals and an ability to make collective decisions that favor the common good. They mirror most of our familiar behaviors.

And yet they do all those things in stunningly different ways from humans, getting to what look like the same destinations without any of the same highway systems or modes of transport. That reflection we recognize is eerily superficial, because what drives the behaviors is not what drives our own. Underneath the maternal care, the language, the system of social favors given and returned is a handful of nerve cells casually strung together in a few small clusters along the body wall. No cerebrum, no right and left hemispheres, not even that so-called reptilian brain part, the cerebellum. They don't have a pituitary gland, or a system of hormones like ours. And yet a sphecid wasp with a body smaller than a kidney bean can dig a burrow in the sand, go off to find a caterpillar just the right size to feed her young, and bring it back to the burrow, remembering where it was and how many other caterpillars she had already brought there. Most of us couldn't find a single caterpillar if we were commanded to do so, much less bring it back to a site the equivalent of a county away. A whole ant colony, with all the drama of the queen suppressing the reproduction of her daughters, can live inside an acorn. A female insect can survey an array of frantically displaying males, select one on the basis of a tiny difference in song, color, or smell, and then store his sperm for weeks or even years before selectively using a particular mate's DNA to fertilize some — and only some — of her eggs.

How is that possible? How can you get what looks like human reasoning, even human love, when you lack not only a human brain but even the chemicals in the blood that drive human emotions? It is easy to endow a fellow warm-blooded creature, for example, a dog or a bird, with motivations and feelings like our own, harder to do so when the entire nervous system of a fruit fly producing a wing-fluttering courtship song of come-hither would fit on a sesame seed.

Insects bring home the uneasy truth that you don't need a big brain to do big things, and that in turn makes us question how the mind and, dare to say it, the spirit, are related to the brain. It even makes us question what it means to be human. What does it mean to have complex behavior? Does it mean you are smart? Is the complexity of a honeybee nest with its exquisitely economical hexagons equal to that of a Park Avenue brownstone? We all have our prejudices, and even scientists can be terribly vertebrate centric about understanding behavior. A huge fuss is made about the behavioral flexibility it takes for a New Caledonian crow to construct a tool from a leaf to poke a grub out of a branch, or a chimp to use a stick to get termites from a hole in the ground. It's that flexibility, we say, that's important — humans and a few other anointed species can change what they do to suit changing circumstances. We aren't little automatons; we are unique individuals. Behavioral flexibility is taken as the hallmark of intelligence and hence the key to human evolution. It is often linked to brain size, and that in turn is said to be important for allowing our complex behavior.

Natural selection can produce what looks uncannily like intelligent thought or emotion but is no more than the relentless culling of minute variations in genetic makeup, generation after generation, for millions of years. Not only that, but insects too have small personalities, with some showing boldness in new situations and some hanging back with what looks an awful lot like shyness. It's turning out that we haven't cornered the market on individuality, either.

Insects make us question virtually every assumption we have about what makes humans human. They lay bare the workings of evolution.

Insects Are a Window

INSTEAD of a mirror, sometimes insects hold up a window, so that we can see through it and imagine life with different ground rules. Insects wear their skeletons on the outside, and they insouciantly transform from egg to grub to gleaming adult in the space of days. Insects use their antennae to smell and hear in ways we cannot even begin to comprehend, with male moths detecting the odor of a receptive female from a single molecule released miles away. Some bees and butterflies can see in the ultraviolet range, giving them an array of colors we don't have names for. Although, as I discuss in a later chapter, insects can learn more than we have previously given them credit for, they produce their complicated behaviors by and large de novo, without benefit of experience or schooling.

All of that difference means that we can learn from insects without having to claim kinship so insistently, the way we do with the feathered and furred. As the famous evolutionist Richard Dawkins said in an article about the intelligent design controversy, "Many people cannot bear to think that they are cousins not just of chimpanzees and monkeys, but of tapeworms, spiders, and bacteria." This unwillingness is particularly true for insects; it may seem improbable to imagine oneself related to microbes, but it does not offend. But to me that lack of identification with insects is precisely why we can look to them to gain insight into our own lives—we simply cannot anthropomorphize them into cute caricatures of humans.

Our inability to identify with insects can thus help keep us—and them—out of trouble, because we do not insist on making them into what they are not. Primates in particular, and especially chimpanzees, seem so much like little people that we almost cannot believe they are animals. When a pet chimp named Travis attacked a woman in Stamford, Connecticut, in 2009, people were shocked, mouthing, as Charles Siebert in the *New York Times* pointed out, many of the same platitudes as when the proverbially quiet neighbor goes on a murderous rampage. "He seemed so pleasant and mild-mannered." Siebert goes on to note, "There is something about chimpanzees — their tantalizing closeness to us in both appearance and genetic detail — that has always driven human beings to behavioral extremes, actions that reflect a deep discomfort with our own animality, and invariably turn out bad for both us and them."

We don't have the same problems with insects. They are so hard to anthropomorphize, and yet they still have that superficial similarity to us. They challenge us to find an explanation for a behavior without resorting to human-specific quirks of physiology or genetics. Insects allow us to study phenomena — the effect of personality type on health, say — without the confounding factor of the mechanism behind them. In other words, if being hard-driving makes people and rats more likely to die early, you don't know if it's because of the stress itself or because of a hormone such as cortisol that happens to be linked to stress in both cases. But if being hard-driving kills off both people and ants, there must be something in the stress itself that is responsible, because ants don't have the same hormones, or indeed virtually any of the same mechanisms for getting from the environment to the behavior, that people do.

I have rarely if ever found insects frightening, at least in the abstract. But I certainly find them unsettling, reminders of another world. I am in good company; Charles Darwin, in his recounting of his observations of tropical insects, found that the possibility of finding so many different species "is sufficient to disturb the composure of an entomologist's mind, to look forward to the future dimensions of a complete catalogue." Some of that is the lack of expression, of what the psychologists call affect, the outward manifestation of one's inner being. The great entomologist Vincent Dethier, who wrote eloquently about the smallest details of fly behavior, felt that the lack of expression in insects stood in the way of our empathy with them: "One empathizes less, if at all," he said, "with a beetle or a fly which has a comparatively immobile head than with a praying mantis that turns her head and stares at one." That lack of empathy is not a hindrance, at least in my mind, but a help. Dethier also said, "It may be, as Alexander Pope averred, that the proper study of mankind is man; nevertheless, to know in which respects a fly is not a man cannot help but reveal something about ourselves." The newest discoveries about personalities in insects are revealing that connection.

Insects are starting to answer the question of "What does it take?"—to have a personality, to learn, to teach others, to change the world around them—with the humbling and perplexing answer, "Not much." Humbling because they do these things with brains the size of a pinhead, and perplexing because if that's all it takes, what does that mean for us, with our gigantic forebrains and exhaustingly long periods of childhood dependency?

Insects Are Essential

> If all mankind were to disappear, the world would regenerate back to the rich state of equilibrium that existed ten thousand years ago. If insects were to vanish, the environment would collapse into chaos.
>
> — E. O. WILSON

WE ALSO keep coming back to insects because they are, however we may feel about them, extraordinarily important to the earth's functioning as well as our own. Insects help aerate the soil by burrowing through it, and nourish it by leaving their droppings. They eat dead plants and animals that otherwise would clutter up the planet, and release the nutrients back to the soil. They control populations of other invertebrates and vertebrates alike, by eating them or their food or by making them sick. In turn, insects provide food for other organisms. Perhaps most critical, insects are key pollinators of commercial and wild plants alike. All of these activities are performed to some extent by animals other than insects, of course, but the sheer magnitude of insect numbers means that they could not be eliminated without leaving a hole so large that, as Wilson says, the rest of the world's organisms would be unable to continue their lives.

To make the worth of these ecological services, as they are called by scientists, more concrete, in 2006 John Losey from Cornell University and Mace Vaughan of the Xerces Society for Invertebrate Conservation calculated the economic value of four crucial tasks performed by insects: pollination, recreation, dung burial, and pest control of animals that eat crops, including other insects. They chose these categories because of the availability of data, not because of their perceived "importance," and acknowledge that the amount is almost certainly a conservative estimate. The total bill? Over $57 billion in the United States alone, and that just includes so-called wild insects, not domesticated honeybees or silkworms or other species that are reared commercially by people.

The recreational aspect of insects is not, as you might initially think, due to people wandering around the countryside collecting butterflies to be pinned under glass. Instead, Losey and Vaughan examined the importance of insects to hunting, fishing, and wildlife observation, including bird-watching. Fish need to eat insects, and we use insects to catch them. Game birds such as grouse and pheasants rely on insects as food, as do waterfowl such as geese and ducks. And without grubs, flies, and beetles, all those lovely harbingers of spring — the warblers and flycatchers, woodpeckers, and swifts — would perish.

Dung removal is probably not a service to which people give much thought, but our own sewage issues aside, everyone produces waste, as the children's book notes rather more colloquially, and it has to go somewhere. If it weren't for insects, that waste would just linger on the surface of the soil or in the water, tying up nitrogen that could be enriching the soil, and providing a breeding ground for disease-causing organisms. Cattle also tend to shun grass that has been sullied by dung. By burying manure underground, dung beetles come to the rescue in many parts of the world, including the United States. They were introduced into Australia, where they do not occur naturally, to help process the massive quantities of dung produced by the cattle brought to that continent in the late eighteenth century. A friend of mine in Perth, Western Australia, worked with the Dung Beetle Crusade, a campaign sponsored by the government to help deal with the problem, and would take buckets of the beetles around the country.

Pollination deserves a special mention, both because of its importance and because the recent decline in honeybee colonies makes the topic particularly timely. More than 218,000 of the world's 250,000 flowering plants, including 80 percent of the world's species of food plants, rely on pollinators, mainly insects, for reproduction. Losey and Vaughan cite a 1976 publication estimating that 15 to 30 percent of our diet in the United States relies on food sources requiring animal pollinators. In a typical fast-food meal of a hamburger, fries, and a milkshake, most of the components required an insect somewhere along the way; although the wheat in the bun is wind-pollinated, the other plants, from the cucumber for the pickle to the feed eaten by the cow, are insect-pollinated. Nicola Gallai from the University of Montpellier in France and her colleagues estimated the world economic value of pollination to be $153 billion, pointing out that this is nearly 10 percent of the value of agricultural production used for human food in 2005. Even more graphically, researchers with the Forgotten Pollinators Campaign in Arizona calculated that one in every three bites of food is made possible by a pollinator. We tend to think primarily about honeybees when it comes to pollination, but hundreds of bee and other insect species help pollinate crops, including the blue orchard bee, the southeastern blueberry bee, and the squash bee. Bees are about much more than honey.

Insects Are Hidden

DESPITE all of the aforementioned virtues, it is undeniable that insects will never fall into the category of what biologists call "charismatic megafauna," the large showy animals such as elephants and eagles that attract the attention of the public and help make the case for conservationists. When whales are endangered, people want to pass legislation and protest in storm-tossed boats. When a butterfly is endangered, people chuckle, and that's if they are feeling sympathetic. In the part of southern California where I live, endangered species are political footballs. Multimillion dollar housing developments can hinge on endangered species occurring on the land where they are planned, and when the Delhi Sands flower-loving fly was put on the list, people were not exactly imagining their wingbeats pulsing over the dunes as stirring music played, the way they would if the species in question were an eagle. It wasn't just that the flies were, well, flies, and hence lumped with vermin, it was that they were invisible to virtually everyone. Why should we save something we'd never even seen?

Yet this seemingly innocuous, easily overlooked quality of insects, belying the extraordinary activity going on under our noses, is exactly what draws those of us in the know to them. In 1991, the Society for the Study of Evolution held its annual meeting in Hilo, on the Big Island of Hawaii. I wanted to go for the usual reasons one goes to scientific get-togethers: people would give talks on their most recent work, I could meet up with old friends and colleagues, and I could recruit new graduate students or collaborators. Besides, I had never been to Hawaii, and I was also excited about seeing the sights, from volcanoes to birds that lived nowhere else.

I therefore decided to go a bit early to the Big Island, and entertain myself for a week or so before the meeting started. I have been studying crickets and their parasites since graduate school, and so it seemed obvious, at least at the time, that the entertainment would involve doing something with crickets in Hawaii. A colleague who had done postdoctoral work at the University of Hawaii in Hilo mentioned that an introduced cricket species, *Teleogryllus oceanicus,* was abundant on the lawns and vacant lots around the campus, and so I decided to collect some of them and dissect them to look for parasites. I now wonder just why this seemed to be the inevitable, or at least the best, option as a recreational activity, but regardless, the week before the conference found me and my long-suffering husband standing on the lawn near the university library, wearing headlamps and watching for crickets in the dark.

Crickets are usually rather secretive animals, with the males staying hidden in burrows or leaf litter while they produce their melodic songs. But here, we kept seeing males out walking around on the surface of the grass, brazen as could be, and what was more, they weren't calling. Since calling is the only way male crickets can attract a mate, and since attracting a mate is a cricket's — any insect's — raison d'être, I was puzzled. What were the silent males doing?

In what has turned out to be the only time in my life that I have impressed my husband, also a biologist, with my scientific acumen, I said to him, "The only place I can remember hearing about crickets doing this is in Texas, where they get these acoustically orienting parasitic flies. But I've never heard of any crickets here getting them. I suppose I should look."

As you can probably guess, the next day I was dissecting the previous night's catch of crickets when a white maggot popped out of the body cavity of one of them, like a ghoulish jack-in-the-box. A little more work established that indeed, the crickets in Hilo — and, as it turned out, on Kauai and Oahu as well — didn't only attract the attention of amorous females when they called. They also risked being discovered by flies that use the chirps in a much more sinister way. Once a female fly locates a calling cricket, she deposits tiny larvae on him. A larva, usually one but sometimes two or even three, burrows inside the cricket's body and starts, ever so slowly, to eat his flesh while he is still alive. First it feeds on his body fat, but eventually, as the fly maggot grows until it occupies the entire body, from head to abdomen, it consumes the male's other organs, so that he is a shell that looks like a cricket but is pulsing inside with fly.

I am interested in this grisly process for many reasons, but mainly because it exquisitely illustrates an evolutionary conflict for the males: it is terribly dangerous to call, because males risk attracting the attention of the flies, but calling is the only way to attract a mate. That week in Hilo got me started on a research program I have continued ever since, trying to discover how evolution has worked out the crickets' dilemma. We work, of course, at night, when most of the locals as well as the tourists are elsewhere, in places that tourists would never think to go, and watch as the drama unfolds in the grass. I have learned a great deal from the crickets, and the whole time I feel as if I am in possession of an enormous secret that no one else in the islands, as they drink mai tais and lie on the beach, has any idea exists.

I am fully aware that from most people's perspectives, that's exactly as it should be, and that knowing about pale sticky maggots bursting Alien-like out of the living bodies of other organisms wouldn't enhance their Hawaiian experience one bit. But for some of us, that sense of being in on a hidden world is exactly why we remain fascinated by insects. Several years into the project, I brought my graduate student Robin to Hawaii to study the crickets, and on her first trip we set out a trap to catch some of the flies, a relatively easy matter because of their single-minded attentiveness to the sound of a cricket. All we had to do was play cricket song through a speaker with a tile placed in front of it; the tile was covered with a sticky substance so the flies couldn't get away once they had been attracted to the song.

We turned on the recording and sat on a bench several yards away. After about twenty minutes I told Robin to go check the tile. She came sprinting back, visibly excited. More than a dozen flies speckled the tile, their wings buzzing in frustration. But Robin wasn't just satisfied at a successful experiment; she was also taken aback. The flies are not insubstantial, being about the size of a small housefly. But she'd never seen one before. Where, she wanted to know, had they been all this time?

It's simple, I responded. You've never seen them because you don't have anything that they want. But now you know they are there, and what they are doing. And things will never be the same.

It is exactly this feeling of a mysterious intricate drama being played out under our noses while most people remain unaware of its existence that makes us keep wanting to understand the lives of insects. Their stories seem unbelievable, with each life cycle, each mating ritual, more extraordinary than the last, and yet they are true. The rest of this book goes to places most people never see, as scientists uncover their secrets with techniques as new as proteomics and as old as a nose buried in the grass watching the bugs go by. We are changing our minds about what it takes to learn, about the nature of individuality, and about what a gene really does, all because of insects and the way they both reveal and reflect our own lives.

Authors write fiction about parallel universes, they ponder the possibility of supernatural beings, maybe even the spirits of the departed, traveling in our midst. The ability to glimpse another world is always touted as an allure for those who dabble in the paranormal. But who needs to be able to see dead people when you can see live insects?

per. This it was unable to do, instead tapping fruitlessly at the paper it was completely capable of cutting through. Two barriers are never found in nature, and the bee couldn't perform acts outside its repertoire; we now know that it lacks the kind of neurological GPS ("if a new roadblock appears, repeat steps A through G until you see air") necessary to adapt to altered circumstances. Fabre tut-tutted over the bee's ineptitude, noting, "The insect would have to repeat the act which it has just accomplished, the act which it is not intended to perform more than once in its life; it would, in short, have to make into a double act that which by nature is a single one; and the insect cannot do this, for the sole reason that it has not the wish to. The Mason-bee perishes for lack of the smallest gleam of intelligence."

Later scientists were equally condescending, noting with belittling superiority that although quite a few kinds of insects can perform remarkable tasks, they cannot learn from experience the way we humans can. In the late nineteenth century, the English physician David Douglas Cunningham was posted to the Indian Medical Service in Calcutta, where in addition to studying the pathology of infectious diseases he made detailed observations of the local flora and fauna, including the many large and easily observed insects. He was prepared to admit that some of the large wasps that provisioned their young with paralyzed caterpillars and other prey possessed something along the lines of what he termed *intellect,* given their complex behavior. But he was also fond of performing "practical jests" on the wasps. The females built mud nests on many objects, including the pipes in his study, and Lieutenant Colonel Cunningham enjoyed occasionally moving the pipe a foot or two from its original location while a wasp was out foraging for prey. He noted that it was then "amusing to observe the astonishment of its tenant when she returns to find her nest gone, and wanders round in perplexity until it is replaced and joyfully recognized." One could certainly wonder about how hard up for amusement one has to be before taking up playing jokes on wasps, but regardless, the same note of self-satisfaction creeps into Cunningham's writing that is seen in the writings of most of the early naturalists. Not being able to find something after it was moved, or being unable to recognize a novel feature in the environment must mean that insects, regardless of their awe-inspiring abilities to construct elaborate hives and find flowers miles away, are dimwitted at heart.

But in fact, it is turning out that here too our faith in our uniqueness may be misplaced, and that insects are capable of feats of intelligence that qualitatively, at least, may be quite similar to our own. This finding has many useful implications, from the construction of better computers and robots to a potential cure for brain damage. And it also challenges our ideas about what our own enormous brains might be for.

Six-Legged Smarts

THE LIKELIEST candidates for insect intelligence, or at least the first ones to be considered by naturalists, have always been the bees, wasps, and ants. Partly this is because we see them more — in our gardens and kitchens — and they seem to be doing things, such as finding food and taking it back to their nest or hive, that require something resembling reasoning. Partly it is because of the sociability of many species, since we use our own intelligence to interact with each other so much. And partly, I think, it has something to do with the way that such insects use objects in their environment, whether it is to build paper cells from chewed wood pulp or to remove pollen from flowers and cram it into the built-in shopping bags on a bee's leg. Animals that have possessions seem smarter, somehow, which may be a comment on our own valuing of material goods.

Fabre, Cunningham, and a host of other naturalists paid particular attention to the provisioning wasps and bees. These relatives of yellow jackets and honeybees do not live in social groups with a queen and workers. Instead, once she has mated, a single female searches for prey such as caterpillars or large toothsome spiders. After capturing the item, she stings it so that it is paralyzed but not dead, a kind of suspended animation refrigeration system. She lugs her victim to her nest, which may be a burrow in the soil or a custom-built cell on the surface of an object, as with the pipe-loving butts of Cunningham's "jests," and lays an egg on it. After the egg hatches, the young larva has a ready food supply that won't spoil. Depending on the species, the mother may return many times to add prey to supplement the larder or to lay more eggs in additional chambers.

While grisly in certain respects, the wasp's behavior undeniably requires two of the prerequisites for intelligence: learning and memory. The mother wasp has to remember where her burrow is, find the correct size and number of prey — in one species, the number of food items brought back to the nest is calibrated to the needs of the hungry waiting larvae — and go back to the correct place. All of this cannot be done purely by rote, because each nest is built anew, each cell provisioned separately, and each prey item puts up a different fight. The wasps seem to use landmarks to find their nests, like remembering where one's house is by recalling the location of the Starbucks at the corner, and if the landmarks are moved, the wasps fly around the area, like the agitated subjects of the jokes played by Cunningham. In their defense, incidentally, one wonders how most of us would do if we suddenly found the aforementioned coffee shop lifted in its entirety off of the block, between one latte and the next, and whether we too wouldn't mill around the area, unable to believe our eyes.

Even more impressive than the ability of these wasps is that of another species of wasp that exploits the provisioning kind. These parasites do not care about the wasp larvae waiting for their paralyzed meal, but about the caterpillars that are brought to the larder. Instead of going out and hunting down their own prey, the parasitic wasps capitalize on the food brought in by the hunters and lay their own eggs on the item. The problem is that only a very narrow window of opportunity to lay an egg on the caterpillar exists, which is during the time that the caterpillar is being dragged into the nest by the wasp that first captured it. So instead of trusting to luck to find a host at exactly the right moment, the parasitic wasp performs a reconnaissance mission, flying around areas where the provisioning wasps are likely to be digging their nests, an activity that takes quite a long time and is much more apparent than the brief provisioning period. Once a nest-building wasp is detected, the parasitic wasp remembers where the nest is located and keeps that nest site under surveillance, so that she can spot when provisioning occurs, often many days later. Then she slips in and hurriedly lays her own eggs on the caterpillar.

Yet another species of parasitic wasp lays its eggs on clusters of checkerspot butterfly eggs. The catch here is that the eggs can be successfully parasitized only for the few hours when the checkerspot babies have developed into first-stage larvae but have not yet broken out of the egg. The wasp circumvents this difficulty by learning where the eggs are ahead of time and then monitoring their progress until they are ready, with some individual wasps finding an egg cluster and then revisiting it for up to three weeks, a substantial portion of the wasp's lifetime.

The wasps and their relatives among the other social insects are not the only ones that can learn new things. The caterpillars and butterflies the wasps use as prey are also capable of learning, and they can also develop preferences for particular foods, depending on the type of plant on which their mother laid her eggs. Such food snobbery is of more than academic interest, since some pest caterpillars that eat crops, for example, the young of the familiar cabbage white butterfly, can learn to eat new varieties of cruciferous vegetables; planting broccoli in hopes of evading butterflies that grew up eating cauliflower is futile. Interestingly, not all kinds of butterflies can learn to go to one kind of plant rather than another; checkerspots, eastern swallowtails, and a species of *Heliconius* butterfly all seem to be relative dullards. You can rear them on one kind of plant, but if you try to train them to visit another kind when it's time to lay eggs, the mother butterflies just won't make the switch. Perhaps it's not stupidity so much as brand loyalty, like refusing to accept Pepsi instead of Coke even if the former is on sale.

Parents often swear that their children are born picky eaters, and that they cannot be taught to prefer healthy snacks. But grasshoppers and their relatives the locusts can be taught to determine the nutritional content of different plants and feed preferentially on the most nourishing ones. In the laboratory, grasshoppers can be fed little cubes of synthetic diet, kind of like the power gels consumed by marathon runners, and the contents of the cubes varied according to the experiment. In one study, groups of locusts were given food lacking either protein or digestible carbohydrates. The experimenters gave one food in a yellow tube and one in a green tube, alternating the association between subjects, and then let the insects feed on a balanced diet for a few days to make sure they didn't become malnourished. Then, the locusts were deprived of food for four hours, a rather long time between meals for the insects, which usually eat more or less nonstop. When the locusts were placed in a test chamber containing yellow and green tubes, but no food, they went to the color associated with the nutrient—either protein or carbs—they had been lacking. This feat is particularly impressive because it isn't just the grasshoppers having some holistic instinct for eating what is good for them, but a learned association between color and a nutritional deficit. Toddlers, take note. Admittedly, the researchers didn't try offering the insects a choice between the hopper equivalent of Twinkies and that of tofu, but then I am not sure quite how one would go about determining what insect junk food would be like.

Honeybees have long been known to navigate using landmarks and use information from each other to find food, as I discuss below and in another chapter, but a recently discovered ability deserves special mention: they can count. The ability to enumerate objects is considered one of those gold standards of intelligence by scientists, and several kinds of primates, some other mammals such as dolphins and dogs, and psychologist Irene Pepperberg's late African gray parrot, Alex, have been shown to do so. Still, you just don't think about insects in the same breath as you do arithmetic. But scientists Marie Dacke and Mandyam Srinivasan of the Australian National University in Canberra trained the bees to fly down a tunnel toward a food reward, using landmarks set along the walls and floor. To get to the food, the bees couldn't simply memorize the position of the landmarks, because the locations of the landmarks were shifted every 5 minutes. Instead, the bees had to learn that the food could be found at the base of landmark number 1, 2, 3, 4, or 5, depending on the individual experiment. Counting to four was mastered relatively easily, but getting to five proved challenging. Nonetheless, that the bees could generalize to a number at all, rather than simply flying until they saw an object in the same place it had been before, is an extraordinary accomplishment.

The bees' ability is exciting not only because it helps demolish that boundary of the backbone with regard to intelligence, but because being forced to design the experiments required to demonstrate counting in a creature so different from us makes us strip down our methods to their essentials. Finding out if your three-year-old can count is one thing. But how do you come up with a test for counting, or learning in general, when your subjects can't talk, walk on two legs, point to anything, or even get rewarded with something they want, the way most people can? If we can design ways to study animals with these limitations, maybe it will help us work more effectively to test humans with limited abilities, or even design computer programs that could substitute for the abilities that are lacking.

Figuring out exactly how to test insect intelligence in a way that is meaningful to them but also tells us something is challenging. Reuven Dukas, a biologist at McMaster University in Canada, has studied learning in a wide variety of insects and thinks we may only be scratching the surface of their abilities. After all, if insects don't learn something, he says, echoing teachers everywhere, "Is it because I'm not a good teacher or because the animal doesn't learn?" It's always hard to know what tasks an animal will be able to perform that we can then generalize to other species. Jan Wessnitzer and colleagues from the University of Edinburgh showed that my favorite insects, crickets, could relocate a particular spot on the floor using objects in a photo along the wall of their experimental arena as landmarks — the best navigational aid was a rather stark landscape that looked like a desert in the American Southwest. The training scheme they used consisted of a floor heated to an uncomfortable temperature except for a single cooler spot that the crickets presumably preferred to stand on. It was called, without comment, the Tennessee Williams paradigm.

The Face Is Familiar, but What about the Antennae?

LEARNING about food sources is one thing, since it is a natural behavior on the part of many insects, perhaps particularly honeybees. But scientists are now demonstrating that insects can be taught far more sophisticated tasks, sometimes having no apparent relation to their day-to-day requirements. Recently, for example, Shaowu Zhang and his colleagues trained honeybees to be extraordinarily discriminating in their decision making. The bees were given a reward if they chose a particular pattern on a card, but the "right" choice depended on whether it was morning or afternoon, whether the bees were out visiting flowers or returning to the hive, or a combination of both. The bees took a while to learn their task, but they eventually could make the distinctions, an impressive cognitive feat. Bees from laboratories in both Australia and Germany were tested, and in a happy blow for global diplomacy, turned out to be roughly equal at the task.

But remembering to choose one visual cue over another pales in comparison to another bee achievement: bees can learn to recognize individual human faces. Adrian Dyer at La Trobe University in Melbourne, Australia, and his colleagues there and at Cambridge University in England rewarded honeybees with a sip of sugar solution if they flew toward a particular image, a technique that has frequently been used by researchers. What was novel was the kind of image in his experiments: a black-and-white photograph of a man from a stock collection, compared with a photo of a different person, the same face upside down, and a drawing. Not all the bees got it right, but those that did could remember an individual face several days after their initial training. Dyer isn't suggesting that the bees actually "know" what they are looking at, or that they spend their days scrutinizing the people around them or developing an attachment to the beekeeper. They can't possibly undergo the same cognitive processes that we do when we recognize each other, given their limited nervous systems. Instead, Dyer believes that the ability is probably related to their skill at distinguishing one flower from another while foraging, something more useful in a bee's life. In other words, a bee that can tell a columbine from a daisy could use the same technique to tell a Roman-nosed individual from a snub-nosed one. Dyer went on to demonstrate that honeybees could discriminate among photographs of very similar natural scenes, with images of forests that differed only in the orientation of the branches, an ability that probably makes returning to the hive after a long foraging flight easier to accomplish.

Regardless of how or why they do it, the bees' capacity to learn to recognize human faces has some important implications. Facial recognition has always been one of those skills thought to require a large brain, and psychologists had even speculated that a special part of the human brain is devoted to just that task. But bees don't have any of the same brain components that humans and other vertebrates do, so such a specialized structure must not be necessary to accomplish the discrimination. As Mandyam Srinivasan said, "Sometimes I wonder what we are doing with two-kilogram brains."

In addition to further blurring those boundaries between human and insect, there are some practical uses for the discovery.

Computerized facial recognition would be a boon to security and crime-fighting agencies, and studying the mechanisms behind the bees' ability might yield insights into how to create such programs. I was seized by the image of a chamber with a bee at airport security, for instance, scrutinizing the faces of passengers to look for matches with photos of known terrorists. Whether this would work better than some of the current efforts is an interesting question.

Some humans themselves cannot distinguish among human faces, a condition known as prosopagnosia, or face blindness, thought to be due to a genetic defect; one estimate claims that 2.5 percent of the population suffers from some form of it. Some people with prosopagnosia can distinguish individual animals, but not people; Jane Goodall is said to have this form of the disorder. Prosopagnosia can also be present to greater or lesser degrees, so that one can have the disorder under certain circumstances but not others. In severe cases, sufferers cannot recognize their own face in a photograph. It seems to be related to the inability to navigate in the environment, which means that bees might be particularly suitable for using as models for studying the disorder, since of course bees are superstars at locating food sources and remembering nest sites. At the moment, no one has worked out the mechanisms by which the bees learn faces, but if they are linked to the ways in which the bees orient in the wild, understanding the bees' abilities could help people overcome their own face blindness.

Each Ant, Teach Ant

WE—AND other animals—can learn things from objects in the world around us, like Dyer's bees or the wasps that remember the location of a rock near their nest. But most of us remember learning in school, from teachers. Insects may lack classrooms and textbooks, but increasing evidence suggests that they too can learn from, and act as, teachers.

In common use, the word *teaching* usually means the transfer of information from one individual to another. A boy sees his sister feed the dog under the table and promptly learns to get rid of his unwanted broccoli the same way. Under that definition, though, even casual observation of another animal doing something that the observer then does would qualify. You could learn to run away from fires by noting a crowd fleeing a burning building, for instance, but has the crowd actually taught you? Even Charles Darwin suggested that many animals, including insects, do this; bees, he pointed out, could follow another worker flying to a source of nectar. If crickets are placed in a container with other crickets that have been hiding under leaves from predatory spiders, they are more likely to find a shelter and hide themselves. But this kind of use of public information seems a bit too haphazard to be real teaching. Animal psychologists are more stringent in their definition and often require the behavior to happen only when a naïve observer (one that doesn't know how to do the task being taught) is present. That means that although a young male white-crowned sparrow learns his song from his father, the father isn't teaching him, because the adult bird would sing whether or not his son were there. Teaching also has to help the observer while costing the teacher something, usually the time and effort required for the demonstration.

Finding an occurrence of this more narrowly defined behavior in nature has been daunting, and until very recently scientists had essentially no examples of real teaching by animals. Just within the last few years, however, researchers have found three cases of it—one in a bird, one in a mammal (the meerkat), and one, incredibly, in ants. People are often surprised by the selectivity of this group, suggesting that surely some other primate besides humans teaches in a natural setting. At least for the moment, the answer appears to be no, which says something about our anthropocentric desire to only see, or bestow, special qualities on those we think are closest to us. That teaching happens in ants and not monkeys or apes is unsettling for the same reason I love studying insects: it's all about getting to the same destination with different modes of transportation.

As anyone who has had to battle the brown ribbons of workers heading toward the sugar bowl knows, ants follow each other to get to food sources. It looks like they are just marching endlessly, one after the other, perhaps following the smell left behind by earlier foragers, but paying no more attention to each other than riders on the same subway train. Odor does play an important role in leading ants to food. But in at least one ant species, a single worker will actively recruit another ant to follow her to a food source or a new nest, or just to explore a new area, in a process called tandem running. The lead ant goes in front, while the follower keeps contact by tapping her with her antennae. If the follower gets behind, the leader waits for her to catch up, and spends time on the task that wouldn't be needed if the leader were alone, fulfilling the criteria outlined above. According to Ellouise Leadbeater and her Queen Mary University of London colleagues, who didn't do the research but study similar kinds of insect social behavior, "The intimate interaction between leader and follower in a pair of tandemly running ants at first sight bears all the hallmarks of a parent teaching a child to ride a bicycle." After being led, the following ant is able to find the target on her own, showing that she has indeed learned from the leader.

This is big news. As an accomplishment it may not rank with

conveying the beauty of Shakespeare to a high school senior, but it means that even ants can respond to feedback from other individuals and modify their behavior so that they improve their performance. Feedback makes teaching different from so-called telling, where in effect one individual says, "Hey, there's a puddle of jam over in the north corner of the countertop, see you there," and then just takes off for the food. This behavior has therefore made scientists question how they define learning, teaching, and their prerequisites. Some researchers feel that because the ants don't improve the skills of those they teach, but simply lead their students along a path, the behavior doesn't really constitute teaching. But in a paper with the subtitle "Ants Are Sensitive Teachers," Thomas Richardson, who led the original project on tandem running, and his colleagues at the University of Bristol in the United Kingdom muse that the arguments over whether the ants are "really" teaching may just be "tracking our own understanding of what is special when humans teach. . . . We should thereby avoid succumbing to the understandable temptation to use the most exotic, extreme case, i.e., the human one, to define what is perhaps a relatively common phenomenon." In other words, once we find that ants do something like teaching, we should not redefine teaching so only humans can be said to do it. And if ants do teach, what other animals might be showing the same thing, if we only open our minds to see it?

Smarter Is as Smarter Does

THE GENIUS of ants notwithstanding, if the basic components of learning and even intelligence lie within a great many creatures, why then are our minds so different? Why do we talk about crows and raccoons and dolphins being intelligent, but chickens and cows as dumb? Is being smarter always better? And if it is, why haven't all animals evolved to be smarter?

The answers to these questions come from an unlikely source: the humble fruit fly. Now, I can usually sell people on crickets, and ladybugs, ants, and bees already get their own movies, toys, and children's songs. People are less than enthusiastic, though, about the possibility of a sparkling intellect lurking in the sesame seed–sized flies that buzz in clouds around decaying fruit. But in Tad Kawecki's laboratory at the University of Fribourg in Switzerland, fruit flies are contestants in an unending game of Jeopardy, insect style. And some of them are big winners.

The flies don't learn How the West Was Won or Celebrity Children, but they do have to master a category that might be called Distinctive Odors, by deciding whether to feed and then lay their eggs on a substance that smells like orange or one that smells like pineapple. One of the two offerings is infused with quinine, which tastes bitter, and the flies avoid that odor and fly over to the other area. Once the quinine is removed, some of the flies still remember to stay away from the place that had the nasty taste, showing they have truly learned the association. Then Kawecki takes the eggs that were laid in the tasty stuff, rears the adults that emerge, and repeats the whole experiment again and again. This means that only the genes from the flies that performed the discrimination correctly are passed on to the next generation. It's the same principle — artificial selection — that farmers have used for centuries to generate cows that give a lot of milk or corn that has large ears, but much faster and with an end product of faster-learning flies rather than county fair material.

Doing these experiments requires painstaking maintenance of the tiny flies in hundreds of jars held under exactly identical conditions — the same temperature, the same food, and in complete darkness. Most modern biology buildings have elaborate facilities for keeping the insects, but of course many scientists labor in less-than-ideal circumstances. As it happened, Kawecki used to work at the University of Basel, also in Switzerland, where his lab was in a crumbling fifteenth-century building in which the doctoral students used a former lecture hall of Friedrich Nietzsche for their office. Although charmingly located on the banks of the Rhine River and architecturally impressive, the building suffered from a variety of maintenance ills, many of which required the service people to enter the attic. The attic in turn was occupied by numerous pigeons and swifts, and one of the building maintenance workers complained so vociferously about the birds' lice and fleas he supposedly encountered in his effort to repair things that an exterminator was called in. While most people welcome the removal of insects from their homes, in a building where precious experimental flies are being kept, the situation is somewhat different. Kawecki and his colleagues made numerous panicky phone calls to the exterminators to make sure the process wouldn't decimate their subjects, and were assured that all would be well.

Unfortunately, as Kawecki puts it, "the only animals [the exterminator] knew about were cats and budgerigars," and the insecticide proved fatal to some of the carefully reared fruit flies. Luckily, the scientists had to stagger the breeding of the flies because they didn't have enough room to raise them all at once, so they did not lose all of their years of effort. But Kawecki remains nettled at the company, which never admitted any wrongdoing, instead suggesting "it was our fault, keeping those stupid flies rather than cats and budgerigars, as proper Swiss citizens do."

Despite these setbacks, one generation of flies led to another. Through the selective breeding process, the flies rapidly improved their ability to remember which substance was attractive and which was not, and after about twenty generations, Kawecki had flies that could go to the bug equivalent of Harvard or Princeton. Instead of taking three hours to learn which substance has quinine in it, the new and improved flies knocked the task out in less than an hour. What's more, they could generalize their ability to other tasks that required them to avoid or prefer one odor to another, and even to other stimuli besides odor, which means that the flies were not simply evolving better discrimination of pineapple versus orange, they were actually getting smarter.

Presumably, being able to detect good places to feed and lay eggs faster would also be useful in the real world, outside Kawecki's lab. So why don't flies show this brainiac capacity naturally? To put it another way, if the flies can get to be so smart, why aren't they rich, or at least more successful?

The answer seems to be that they don't live long enough. The life span of flies from the smarter lines averaged 15 percent shorter than their unselected relatives. Furthermore, the smarter females laid fewer eggs, an ominous characteristic from the standpoint of evolution, since it means fewer potential copies of genes in future generations. The decreased survival was particularly notable when food was in short supply, which gives a clue to the reason for the finding: learning is costly, and investing brain resources into intelligence may mean that you pay the price somewhere else. More brain, fewer eggs. The trade-off even occurs within the lifetime of a single fly. A group of flies that was trained to associate an odor with a mechanical shock and then deprived of food and water died 4 hours earlier than flies that were exposed to the smell or the shock but didn't have to go through the training, suggesting that something about the process of remembering the association drained the resources of the diligent flies.

Such trade-offs are common among living things, as I discuss in the chapter on personality. Animals that have many young also tend to have smaller babies, whereas species like us that give birth to one or a few offspring at a time generally produce relatively large ones. Here the trade-off seems to be that when natural selection gives a good learner, it takes away a long life. This could happen at two time scales. Within the lifetime of the fly, the energy a fly acquires could go either to helping it survive longer, or to nervous system machinery, but not both. It may be cheap to upgrade the memory in your laptop, but doing so in the brain is going to cost you.

Over many generations, a different process may be at work. Say that a gene makes a fly smart, but because most genes have more than one effect, it also makes the fly vulnerable to starvation, or maybe more susceptible to infections. If being smart is advantageous enough — in Kawecki's lab, it made the difference between reproducing or not — then the gene conferring it will persist in the population, even if it also has some downsides.

Of course, it's not as if all animals get to go to some primordial retail smorgasbord and shop for a certain number of abilities, with some picking learning, long legs, and a mean tennis serve, while others choose curly eyelashes and a talent for languages but end up dimwitted. Exactly which abilities end up having to trade off against which others is still a mystery. But Kawecki's work suggests that the ability to learn, and hence perhaps intelligence, exacts a high price. And that in turn could shed some light on our own evolution. Humans may well have given up some other abilities when we evolved our large brains. What's more, having to learn everything from infancy, rather than being born with our skills, makes our childhoods vulnerable to everything from hot stoves to saber-toothed tigers or their modern-day equivalents. The trade-off in our case must have been substantial, but scientists are still wondering about exactly what it was that we humans had to pay for our intellect.

Better Learning through Chemistry

ONE OF the wonderful things about using animals such as fruit flies and other insects to study learning is that they present a window into the brain. Exactly what happens in the body when you learn the capital of Mongolia, or how to get to the theater? We all have some vague idea that nerve cells send messages somewhere, that electrical impulses in the brain do . . . something. And we can use complicated brain scans with colorful images of different centers of nerve activity, or detailed dissections, to try and figure out what that might be. But insects, unlike humans, let us alter a chemical here, or breed up offspring with a special mutation there, which means it is sometimes possible to pinpoint precisely what makes an individual able to perform a certain task. If one bug has gene variant A, and another bug's genes are exactly like it except for having gene variant B, and if the two differ in the time it takes them to find a food reward in a maze, then presto, we have a gene linked to learning.

In most cases, those carefully bred and engineered insects are fruit flies. In the chapter on personality I mention the "rover" and "sitter" flies, which exhibit genetically programmed differences in behavior. Kawecki and his colleagues, most notably Frederic Mery, examined these tendencies in light of their studies on learning. Each behavior is associated with a form of a single gene, and flies with the rover variant of the rover-sitter gene have better short-term, but worse long-term memory, something of a reversed Alzheimer's, where it is easier for sufferers to recall events of decades past than what they had for lunch. Sitters show the opposite pattern and can remember associations from several days ago, but not the fly equivalent of what they ate at an earlier meal. These different strengths and weaknesses make sense in the natural world of a fly; rovers are likely to move from one food source to another, so being able to quickly learn whether a given fruit is ripe or not is more important than remembering what happened in the more distant past. Together with Marla Sokolowski from the University of Toronto, who first discovered the rover-sitter dichotomy and has worked on its details for many years, the scientists then discovered that the differences in memory can be manipulated by increasing or decreasing the amount of an enzyme in the odor detection centers of the insects' brains. That enzyme may be the key to the trade-off between memory types, at least in flies, and suggests some interesting directions for similar studies in people.

Another set of experiments focused on a different chemical. Using a modified version of the Tennessee Williams paradigm, in which flies are placed into a chamber that heats up on one side when the flies move to it, a group of researchers from the University of Missouri recently demonstrated that serotonin, the same brain chemical that features so prominently in human depression and its treatment, is key to the tiny flies being able to learn to avoid the hot spots.

The ability to stick to a task after having been distracted—something many children with learning disabilities struggle to accomplish—is also controlled by a few nerve cells and chemicals. Flies tend to move toward a visual object, for example, a stripe on the end of their container. If you remove the goal and show them a "distracter" stripe somewhere else, they veer off for a short time but can still remember where the original stripe was located. Geneticists have bred flies with mutations in various genes that produce chemicals important in learning (as with many specialized strains of fruit flies, these have fanciful names such as *dunce* and *ignorant*), and it turns out that while some of the mutants can still perform the task of recalling their goal as well as normal flies, others cannot. The mutant *amnesiac,* for example, learns just fine in the first place, but forgets what it learned almost immediately. This distressing tendency can be attributed to a defect in a single neurochemical, one that is extremely similar to a chemical in the human nervous system. Being able to break down a behavior such as recovering after a distraction into components so fine that we can determine exactly which gene is responsible for which part of learning is possible only in insects, at least so far, but maybe someday we will be able to extend this kind of detailed understanding to our own learning difficulties. What's more, the prospect of altering or curing defects in memory with gene therapy in insects suggests that similar treatments may eventually substitute for drugs or surgery in humans, a solution that could have fewer side effects and be targeted more precisely than current approaches.

He Who Learns Last

WHICH came first, learning or instinct? Because humans rely so heavily on learning, we tend to think of it as an innovation, an evolutionary novelty that we alone have mastered. In effect, we like to think we invented invention. But centuries ago, naturalists believed that instincts, behaviors that are performed more or less the same way every time, arose after learning. Early animals, they claimed, had to learn things from scratch, and then after time, the repetition of a task was somehow impressed into the fiber of the organism so that eventually it became instinctive. This idea was particularly championed by Jean-Baptiste Lamarck, the French biologist whose ideas about the inheritance of acquired characteristics were first embraced by early evolutionists, including Charles Darwin, but later discredited. Knowing nothing about how genes and chromosomes could be passed from parents to offspring, Lamarck and his contemporaries reasoned that if, say, a horselike animal continually reached for its food at the top of a tree, its neck would become longer. This greater development would then somehow be passed onto its offspring, who in turn would develop even longer necks, eventually resulting in what we now call a giraffe.

Lamarck was quite interested in invertebrates, a subject not much studied by naturalists in the late eighteenth and early nineteenth centuries, and he was intrigued by the idea that they exhibited such fixed behaviors. It seemed reasonable to him and many of his contemporaries that doing something, whether following an odor trail or learning to count, could cause a permanent change in the body, and that such changes could be inherited. Of course, we now know that the genes cannot be influenced in exactly the way that Lamarck imagined. And it is likely that both learned and instinctive behaviors evolved together. Most behaviors, even in insects, are due to a combination of influences from the environment, and hence subject to learning, and influences from the genes, and hence instinctive, making the old argument somewhat moot.

Dan Papaj, a biologist at the University of Arizona, doesn't believe Lamarck himself was correct, but he does wonder if there aren't new ways in which learned behavior could influence evolution. He works with a variety of species, from butterflies to parasitic wasps, to see just how learning operates in nature. He points out that the idea that fixed behaviors could have arisen from something an ancestral insect learned to do is not as far-fetched as it might seem. Researchers in the fields of robotics and artificial intelligence are particularly interested in how changes in stimuli—that is, the response a computer gets when it executes an action—could then make the computer's actions more sophisticated. It would be amusing if the behaviors so derided by Fabre and Cunningham turned out to pave the way for better, and more flexible, computers.

Finally, social insects are well known for their genetically hardwired altruism; honeybees can't help committing suicide when they sting an intruder in defense of the colony, because the stinging apparatus remains imbedded in the victim, tearing the innards of the bee asunder after the sting. But it has just come to light that ants, at least, can also choose to rescue their kin from harm even when the peril is novel. Elise Nowbahari and her colleagues in France and the United States took ants, partially submerged them in sand, and restrained them with a nylon filament so that their bindings were concealed under the surface. The scientists then released either strangers or nest mates of the victim and watched the ants' behavior. If, and only if, the entrapped ant was from the same nest, the other ants hurried over, dug her out, and bit the snare away. Ants from foreign colonies, even though they were the same species, were left to struggle helplessly.

Such a complex sequence of behaviors pushes the boundaries of what we thought an insect could learn. And if the same ability applies to species other than ants, we might want to rethink those sticky traps that attract cockroaches and trap them, alive and kicking, on the surface. If the roaches become able to rescue their fellows by nibbling through the glue, you have to start wondering if they might then be capable of plotting revenge.

CHAPTER 2

Six Legs and a Genome

SOME of the most cutting-edge discoveries about insect molecular genetics, and therefore about how genes do and don't dictate complex behavior, have been made because Gene Robinson was tired of harvesting fruit. As a student worker on a kibbutz in Israel, he was asked to "help out with the bees temporarily, and since I was bored to tears picking grapefruits, I volunteered. I remember I was smitten that very first day."

In his correspondence, he glosses over exactly why the bees were so appealing, but despite parental skepticism (he summarizes his mother's response as: "No doctor, no lawyer, where did we go wrong?"), Robinson went on to pursue a master's and later a Ph.D. in entomology. Now at the University of Illinois, he still professes an unabashed love of bees, which he has parlayed into one of the most compelling uses of genomics, the study of an organism's entire mass of DNA, anywhere in the world of biology. Robinson is interested in just how a complicated behavior such as the division of labor in a honeybee colony, where some bees go out and forage among the flowers while others stay home and nurture the young, is derived, first from the hormones coursing through the bee's body, then via the firing of nerve cells in the brain, and ultimately from the minuscule variations within a gene that directs the activity. He calls what he does sociogenomics, the molecular genetics of social behavior. It is where the genetic rubber meets the behavioral road, and it can best be understood using insects.

Before explaining sociogenomics, a bit of background about the new age of genomics, and about what we mean by sequencing a genome or having a "genome project," is in order. The genome is the total set of DNA in an organism, arranged into the chromosomes that are characteristic of each species; humans have twenty-three pairs of chromosomes, while cats have nineteen pairs, cows have thirty, silkworms have twenty-seven or twenty-eight, and a species of ant has just one. Sequencing a genome means determining the order of the four chemical bases that are the building blocks of the helix of DNA. The bases are called adenine, thymine, guanine, and cytosine, usually abbreviated with their initials A, T, G, and C. The genes themselves are particular sequences of the bases that contain instructions on the manufacture of proteins that make up the structure of the body or instructions on regulating when and how other genes become activated. Not all of the DNA consists of genes; scientists knew going into the Human Genome Project, the first of such efforts, that some amount of the material on the chromosomes would be noncoding, meaning it does not contain information about either gene regulation or the making of a protein. The genome sequence therefore consists of a long — a very, very long — string of four letters, grouped together in a particular arrangement unique to each species.

Once the Human Genome Project was completed in 2003, it was clear that more genomes needed sequencing. Many scientists wanted to put two animals next on the list. First would be the zebra fish, as a way to examine genes responsible for the development of a fertilized egg into an adult organism, and then the laboratory mouse, because as a mammal we could more easily compare its genes to those of people. Nobel laureate Sydney Brenner demurred, saying that "the mouse is too close. It hasn't had enough time to randomize, so you are confused by the commonness of origin."

What he means is that because we so recently shared a common ancestor with mice, our genetic material is already very similar to theirs. But which genes are the essential ones, the ones retained through hundreds of millions of years? How have genes changed to perform different functions? To answer that, we need insects. It's been 250 million years since the mosquito *Anopheles gambiae* and the fruit fly *Drosophila melanogaster* shared a common ancestor. That's roughly the same evolutionary distance that exists between humans and fishes, a third more than the distance between humans and chickens.

Of course, it's not an either-or situation. The zebra fish and mouse genomes have now been sequenced, along with those of the chicken, the African clawed frog, and a nematode called *Caenorhabditis elegans.* Genome projects are in progress for a whole host of others, including the European hedgehog, the green anole (a small lizard often sold in pet stores as a chameleon, although it is only distantly related to the true chameleons), and the gorilla, in addition to many invertebrates. Nevertheless, insects can reveal the process of evolution in ways that no other group of organisms can.

As I already pointed out, insects are the most diverse group of organisms on the planet—there are more kinds of insects than any other organism, they live almost anywhere except deep in the ocean, and they vary enormously in size, shape, food habits, and virtually every other aspect of life. A queen ant can live for decades in her nest, while tiny midges that circle over fast-running Appalachian streams can dispatch a whole adult lifetime, complete with finding a partner, mating, and laying eggs, in a prompt 45 minutes. That diversity makes it much easier to answer questions about the genes responsible for traits such as life span or body size, because we have so many different types of animals to compare. Even if we had genomes for all the primates, say, or even all the mammals, it wouldn't be as useful as having genomes for as many types of insects, because compared with insects, one monkey is pretty much the same as another when it comes to appearance and even behavior. A monkey is a lot more like a mouse than a grasshopper is like a flea. And of course insects are important vectors of diseases from malaria to typhus, as well as linchpins of our agriculture through pollination and pests because of their fondness for the same foods we eat. Without them, we cannot understand what makes life tick.

What's more, because we shared a common ancestor with insects so long ago, we can use them as a way to explore how we arrived at similar-seeming destinations with such radically different modes of transportation. For example, we are social and spend time and energy taking care of our young. Honeybees are social and spend time and energy taking care of their young, too. We share a fair proportion of genes with honeybees — but are the genes associated with social behavior the same in both of us? If they are different, how do they get similar results? If they are the same, why did the genes persist through evolutionary time in us and them, but not in thousands of other species?

Size, Junk, and Garbage

BEFORE exploring which insects have had their genomes sequenced and what those sequences tell us, it is necessary to look at a different kind of large-scale genetic information we can get for living things: genome size. Before we knew much about the chemical bases that comprise the DNA inside a cell, we could at least determine the amount of DNA itself. Indeed, calculating an object's size is one of the first things we do with something new, whether that something is a previously undiscovered mountain, a recently incorporated township, or a newborn baby (why the vital statistics of weight and length are so often included on birth announcements is a mystery, at least to me, but it testifies to our obsession with measurement).

Ever since the DNA molecule was discovered in the late 1800s, scientists were interested in the amount of it in different kinds of animals and plants. In the 1940s and early 1950s, the "DNA constancy" hypothesis, which stated that the nuclei of cells in various tissues contained about the same amount of DNA, and that this was roughly twice the amount contained in sperm cells, was used to test, and eventually support, the notion that DNA was indeed the source of genetic material.

Once this idea was accepted, it seemed plausible that the more DNA a species had, the more genes it possessed, and therefore the more complex it could be. Intuitively, people looked at genes like money in the bank; the more you have, the more you can buy. Scientists thus expected that smaller, simpler organisms such as amoebas or flatworms would have less DNA per cell than hamsters or birds of paradise. Much to their surprise, this turns out not to be the case. The amount of DNA — weighed in picograms, or trillionths of a gram — is not related to the apparent complexity of the animal or plant in which it resides. Knowing genome size is useful in deciding which organisms should have their genomes sequenced, for the purely practical reason that sequencing smaller genomes is cheaper.

Animals vary seven-thousand-fold in their genome size, and as you might expect, insects are champions of this variation. Among mammals, the smallest genome of 1.73 picograms resides in the Asian bent-winged bat, while the largest, in the red viscacha rat from South America, is really not all that much bigger, at 8.4 picograms. This size difference is dwarfed by insects, which vary 170-fold in genome size. Here the champion seems to be a mountain grasshopper, with the diminutive Hessian fly as its sparsely endowed counterpart. Humans, by the way, have genomes of a modest 3.5 picograms, which at least weighs in at more than the house fly, though less than that of the grasshopper.

Aside from the kind of *Trivial Pursuit* cum *Guinness Book of World Records* appeal of this kind of information (though, alas, clues about genome sizes are unlikely to come up in crossword puzzles), what does the variation in genome size—and its lack of relationship to the complexity of the organism in which it resides—mean? Obviously, more isn't better. Bluntly put by Ryan Gregory, a biologist at the University of Guelph and one of the world's leading genome size researchers, this decoupling of DNA content and complexity puts paid "the expectation that genomes consist of the genes, all the genes, and nothing but the genes."

So if the genome contains material other than genes, how did that happen? Furthermore, what exactly is that other material, and what is it doing in there? And why do some organisms seem to have so much more of it than others?

The answers to these questions are intertwined. Some of the "extra" material consists of free-floating bits of DNA, sometimes called transposable elements or, more colorfully, selfish DNA. These arise when a sequence of DNA copies itself several times and then just lingers as part of the genome. It is selfish because, à la Richard Dawkins's selfish gene, the elements benefit by making more copies of themselves, but they do not contribute to the functioning of the organism in which they reside. If there is no disadvantage to the organism of harboring them, or even if there is a cost but no means of getting rid of them exists, they will persist, cluttering up the genome and giving us those oddball genome sizes in some species.

Other noncoding DNA is often called junk DNA, which sometimes is used to mean all types of genetic material aside from the genes themselves, but more properly refers to copies of genes that used to be functional but are now obsolete. Like a manual lawnmower with a broken blade that you tuck away in the garage even after you've bought an electric model, the junk DNA clutters up the genome. In a distinction reminiscent of couples squabbling over organizing the closets, some scientists call DNA *junk* if it's not functional at the moment but could be useful at some hypothetical time, like that lawnmower, but *garbage* if it's not functional now and never will be, like — well, maybe it's best not to offer an example here. As with the transposable elements, junk DNA is thought to accumulate because DNA has an inherent tendency to copy itself unless otherwise halted.

Genome size is often, though not always, a reflection of body size, particularly among insects and other invertebrates. And insects that take longer to develop from eggs into adults have larger genomes as well. Another restriction on insect genome size seems to be the way that the species develops — does it go through a metamorphosis with egg, caterpillar, cocoon, and adult stages, like a butterfly, or does each successive stage look like a slightly pumped-up version of the one before, like a grasshopper? The butterfly types seem to have far smaller genomes than the grasshoppers, for reasons that are unknown. Also perplexing is a link between sperm length, which as I discuss further in a later chapter varies enormously among insects, and genome size. And intriguingly, all insects that exhibit social behavior, including not just bees and wasps but termites, as well as cockroaches that take care of their young after hatching, have reduced genomes, despite the vast evolutionary distance between these groups.

I look forward to the solutions to questions about genome variation, but what I like best about the measurements of genome size is the way they make our selves feel so literal, so concrete. Thinking about how many molecules can be crammed into a cell, imagining the adenines and thymines jostling for position, or the helices spooning like lovers in the nucleus, means that we can visualize who we are with startling clarity. Science writer Carl Zimmer titled his book tracing the history of our understanding of the brain and its relation to the mind *Soul Made Flesh,* in reference to the way that we can now see our essence in neurochemicals and gray matter. Thinking about the actual DNA, doled out in infinitesimal picograms in the genome, seems to make that translation even more tangible.

The Sequential Fruit Fly (and Mosquito and Beetle)

THE FIRST insect to have its genome sequenced was, as you might imagine, that sturdy workhorse *Drosophila melanogaster.* This was followed by the honeybee, a suite of other fruit fly species in the genus *Drosophila,* two mosquito species, the silkworm moth, and the tiny beetle that often inhabits the flour canister in your kitchen. More are on the way, and all are helping us understand the action of evolution on humans as well as our six-legged kin.

Let us begin with the fruit flies. *D. melanogaster* is the model species for genetic research, but other species of *Drosophila* lead lives that are both similar and different. Unlike the cosmopolitan

D. melanogaster, D. sechellia, for example, lives only on the Seychelles Archipelago in the Indian Ocean, where it specializes on eating Morinda fruit, from a usually toxic plant. *Drosophila grimshawi* has elaborately patterned wings and is one of the extraordinarily diverse Hawaiian *Drosophila,* occurring only in a handful of remote locations. It is nearly a hundred times bigger than the puny *D. melanogaster.* A close relative of *D. grimshawi, D. mojavensis* is native to the Sonoran desert of the Southwestern United States and breeds on the spiky organ pipe cactus.

The flies were chosen deliberately to cover a broad range of evolutionary history; the different species shared a common ancestor anywhere from *half a million to sixty million years* ago. This is approximately the same distance between humans and lizards, all within a group of flies in the same genus. Many of the genes are similar in all of the species, but others are surprisingly different. Journalist Heidi Ledford referred to the "turmoil" of the genome that is visible only when genes are compared across species; genes appear and disappear, the time and place for them to be switched on and off is altered. Even those stalwarts the sex chromosomes had some surprises; some genes were thought to be expressed only in males because of their position on the X chromosome, but different species with the same gene did not always express it in the same way. The genes used to code for molecules that fight microbes—part of the fruit fly immune system—are much more variable than others, which makes sense given the rapid rate of change of the disease-causing organisms. Genes for detecting odors, crucial to animals that make their living and find their mates on fermenting vegetation, are also diverse. And in some cases, although the regulation of a pathway for making a protein clearly changed, the protein itself was still being produced, suggesting that so-called transcriptional rewiring might be commonplace.

In contrast to the desire to discover universal principles about the operation of genes from the workings of *Drosophila,* the scientists examining the two species of mosquito that have had their genomes sequenced had a much more practical motivation: they wanted to understand species that have such enormous effects on human health. The first species to be sequenced, *Anopheles gambiae,* is the principal vector of malaria in Africa. The second, *Aedes aegypti,* transmits yellow fever, dengue fever, and the less well known chikungunya virus; the latter was responsible for a recent outbreak in countries bordering the Indian Ocean that caused about 250,000 cases of illness and over two hundred deaths. Two U.S. Department of Agriculture entomologists, Jay Evans and Dawn Gundersen-Rindal, note that *Anopheles* was "the first animal to be sequenced, other than ourselves, whose actions have a strong direct impact on human lives." Although *Aedes* has a much larger genome than *Anopheles,* it doesn't encode many more genes, further supporting the idea that even closely related species can differ in the amount of noncoding DNA.

Once the sequence data can be used to identify functional genes, it should be possible to detect which genes are responsible for, say, successful transmission of the microorganisms that carry disease inside the mosquito's body, or for the mosquito's ability to use odor cues in sweat or exhaled breath to find a human to bite. The hope then is to tinker with these genes and breed a mosquito with a gut that is inhospitable to the malaria parasite, or one that cannot smell a delectably pungent victim nearby. Genes that are used to resist the effects of insecticides could similarly be altered to ensure that the mosquitoes remain vulnerable to certain chemicals.

If the fruit flies were sequenced to take advantage of the classic model system for genetics, and mosquitoes were sequenced in hopes of applying the knowledge to curing human disease, the flour beetle *Tribolium castaneum* could be said to have been sequenced because, well, no project in animal biology is complete without including a beetle. More kinds of beetles have been described than any other single group of animals—with over 350,000 species, one-quarter of all of the species of animals in the world is a type of beetle. The scientists who collaborated to sequence the flour beetle genome boast that beetles are "by far the most evolutionarily successful" multicelled organisms, and list, as if the insects were trying out for some kind of all-star reality television show, the many talents found in the group: "Beetles can luminesce, spit defensive liquids, visually and behaviorally mimic bees and wasps, or chemically mimic ants." I am not sure why these particular abilities are showcased, although there is a kind of "animals you might want with you on a desert island" kind of flavor to the selection. Interestingly, the beetles share with that other highly successful insect group, the ants, a lack of flight in day-to-day life; although most beetles can fly if necessary, their lives are mainly spent walking and tunneling on the ground. Whether this sacrifice of fragile wings is the key to their profligacy is not clear.

Tribolium itself is a good choice, among all those hard-shelled crawling candidates. Because it is easy to rear in large numbers in Petri dishes or other small containers, it has already been the subject of other types of genetic studies for many years. It is also an economically important pest in stored grains, which means that discoveries about its genome could reveal genetic Achilles heels to be exploited in its control, an urgent need since up to now it resists all kinds of insecticides that have been used against it.

Despite all the attention paid to the fruit fly *Drosophila* and its kin, it turns out that *Tribolium* is more of an "ur-insect," so to speak, than the fly—in other words, the flour beetle's genes seem to be less specialized and more like that of the ancestor of the entire class of insects than do the *Drosophila* genes. Over 125 groups of genes that the beetle has in common with humans, for example, don't occur in the other insects whose genomes had been sequenced as of 2009, suggesting that *Tribolium* has some pretty basic genetic material. In fact, nearly half of its genes are ancient, with counterpart genes occurring in vertebrates. This primordial nature means that it will be easier to determine how genes have changed through evolutionary time by comparing various groups to the *Tribolium*, and to determine which genes are responsible for general features of insect biology, such as metamorphosis or molting, and which are more idiosyncratic, say, those controlling the ability to make honey.

As is the case for genome size, and for that matter body shape and appearance, the genetic information from insect genome sequences is much more diverse than that obtained from vertebrates. A few constants appear, such as genes associated with detecting odors or those used to produce compounds that fight disease, but others are far more specialized. Silkworms possess about 1,800 genes that aren't seen in mosquitoes or fruit flies, including some used to make silk; although all insects and spiders use silk in some form or another, for spinning cocoons or dropping down from ceilings, the silkworms seem to have some additional genes exclusive to their lineage.

Of course, the first step after sequencing is to find genes with particular functions. Once that is accomplished, the opportunity arises for new, and sometimes diabolical, methods of pest control. Scientists are currently trying to use genetics to make insects pass on the instruments of their own destruction. A gene that is innocuous in the presence of, say, a particular antibiotic, but lethal otherwise, is inserted into an insect. The insect is then reared on a diet containing the antibiotic until it is an adult, when it no longer feeds, and is released into the wild. After the insects with the manipulated genes mate with normal members of the opposite sex, they produce offspring containing the gene—but those offspring are out in nature, where the lethal gene takes effect. Other even more clever methods are in the works.

As with genome size, studies of genome sequences confirm the presence of a hefty amount of noncoding DNA. One researcher refers to it as "dark matter," similar to the science fiction–like invisible stuff of outer space, which conveys both the mysterious nature of the substance and the almost peevish response that its discovery has elicited. We all seemed to have expected Mother Nature to be more thrifty in her allocation of genetic material, maybe saving that extra DNA, like leftovers at dinner. Shouldn't somebody have made another organism out of those bits and pieces of adenine and cytosine? Or maybe we just don't like the idea that it doesn't take many genes to make a whole complicated being; as Ryan Gregory says, "The strikingly low number of genes required to construct even the most complex organism represents one of the most surprising findings to emerge from the analysis of complete genome sequences." Somehow we seem to feel cheated by our own simplicity.

Of course, it's not that we are simple, per se. It's just that, once again, we are reminded that evolution is a tinkerer, using what's at hand to make its products. I like to think of the nuclei of our cells, not as perfectly tuned whirring machines, each gear essential, but as vast echoing warehouses of factories. Entire machines are outdated and useless, left rusted in a corner but never taken away and demolished. Others are jury-rigged out of pieces from older models and newer ones, rattling jerkily through their paces but ultimately manufacturing something useable.

The Social Genome

ALTHOUGH honeybees, like mosquitoes, are enormously important to human well-being, the sequencing of the honeybee genome was heralded not just because it might help us fight the mysterious decline of colonies throughout North America, but because bees are such extraordinarily social animals. Gene Robinson, who eschewed fruit-picking to devote himself to bees, thinks studying their genomes can show us how animals can become so integrated that they are often described as a single superorganism. According to the great biologist and ant lover E. O. Wilson, "If Earth's social organisms are scored by complexity of communication, division of labour, and intensity of group integration, three pinnacles of evolution stand out: humanity, the jellyfish-like siphonophores [creatures such as the Portuguese Man o' War], and a select assemblage of social insect species." Where does this high degree of interdependence come from?

One of the most surprising pieces of news from the honeybee genome project, published in 2006, was the relative paucity of genes associated with defense against diseases, compared with the other insects that have been examined. Given the crowded conditions of your average hive, one might imagine that pathogens would spread faster than colds at a preschool, which should select for highly vigilant immunity. One possible explanation is that the intense social behaviors of the bees, for example, the grooming and licking that individuals are always bestowing on each other, obviate the need for other defense mechanisms. It is also possible that honeybees, domesticated as they have been for thousands of years, will turn out to be an anomaly in this regard, a question that the sequencing of other social insect genomes should help settle. Two other startling results were the small total number of genes in the honeybee genome, and the apparent conservatism in the rate of the genome's evolution, compared with the mosquito *Anopheles* and *Drosophila melanogaster,* so that at least for some groups of genes, bees are more like vertebrates than those other insects. Contrary to what had been believed previously, in fact, bees seem to have arisen quite early in evolutionary history, branching off before the beetles.

Bees do have a lot of genes associated with producing and detecting pheromones, chemicals used to communicate with other individuals, which is not so surprising given their reliance on signaling within the colony, and they have some new genes that are associated with collecting nectar and pollen. But do they have special "sociality genes"? Several years before the honeybee genome sequence was completed, Gene Robinson noted that the difference between highly social insects such as the bees and solitary species such as *Drosophila* was likely to lie not in the creation of entirely new genes, but in changes in the way the same genes were turned on and off, or in the amount of product a given gene made. With some exceptions, this has turned out to be the case. Indeed, Robinson and his postdoc Amy Toth suggest that just as developmental biologists have discovered "modules" in body plans, with wings, legs, and arms produced from similar groups of genes in different animals, behavior can likewise be broken down into building blocks.

One of the most significant elements of insect sociality is the division of labor that Wilson cited above. Unlike other insects, or even virtually any other animal except for a few oddballs such as naked mole rats, in ants, bees, and termites queens do queenly things like produce eggs, males mate, and workers, well, work. Within the workers, different individuals often specialize on particular tasks, for example, going out and collecting food, or cleaning up the hive.

This division, like the stratification of human industrialized societies, allows the colonies to be much more efficient. And the whole idea of sterile individuals that nonetheless labor for the group as a whole is a hallmark of sophisticated social organization. But what determines the destiny of any one individual?

It's arguable whether being a queen in a social insect colony is enviable or not, what with the continual egg laying and never getting outside, but the dogma used to be that queen honeybees were made, not born, via the feeding of royal jelly, a substance produced from glands in the heads of the workers that is given in lesser and greater amounts to different larval females. Adult bees, regardless of their social status, do not eat royal jelly. If you got a lot of royal jelly, the thinking went, you became a queen, while more modest amounts destined you for a short, chaste life among the colony proletariat. In the words of royalbeejelly.net, "Royal Jelly, the queen's food, makes the queen into a bigger animal with superhero powers," which I suppose is true if being capable of laying massive numbers of eggs is viewed as the insect equivalent of making yourself invisible. The association between upward mobility and royal jelly has given rise to a number of claims about the substance's ability to cure everything from asthma to wrinkles, though in a more sober moment surely someone has pointed out that bees suffer from neither.

But now it's turning out that at least for some social insects, you are not only what you eat, you are also the way you were born. In honeybees, different nutrients interact with the genome to switch some developmental pathways on and off, for a much more complex picture than had originally been supposed. In some ant species and at least one kind of termite, females bearing one version of a gene are more likely to be queens, while females with another version end up being workers. A particularly odd version of this genetic influence on caste occurs in harvester ants, in which two genetic lines coexist; queens belong to one line or the other, but workers are a cross between them. If a queen's eggs are fertilized by a male sharing her pedigree, the larvae become queens, but if the father is from the other line, the daughters become workers (recall that sons are produced only from unfertilized eggs, so they don't enter into the calculation). The difference between queens and workers can also be due not to a gene or genes being present or absent but to the regulation of those genes. A recent study of honeybees found at least two thousand genes that were present in both workers and queens were expressed differently in the brains of the two kinds of individuals, further supporting the idea that it's not just what you have but what you do with it — or what it does to you — that counts.

Queens may also specialize, with multiple reproductive females starting a nest together and then divvying up the duties like housemates, so that one goes out and collects food and the other stays home caring for the offspring. Alternatively, in the fire ant common to the southern United States and named for its painful sting, some colonies have one queen and others have two or more. The fatter queens go solo, whereas the burden is shared, literally and figuratively, in the nests ruled by multiple, lighter queens. Queen physiology, and the way the queens are treated by the workers, are both controlled by genes.

The role taken on by a worker had also been thought to be, if not diet related, at least environmentally determined, with all older worker bees, for example, doing more foraging and younger ones staying behind as "nurses." Now, however, the picture seems both more complex and more genetically determined. The age-related changes in tasks occur, to be sure, but altering the genes can change the workers' behavior, making them go out to forage at a younger age than they normally would. At the same time, foraging is influenced by social cues such as the age of other colony members and the type of pheromones given off in the hive, which in turn can feed back to the worker and change the hormones secreted inside the worker's body, further altering behavior. As in the queen-worker distinction, gene expression differs depending on the task the workers do. Hives with differences in genetic makeup also show different patterns of work. Most interesting, when a queen had mated with multiple males, the resulting blended family of workers was more efficient at making honeycomb, rearing the young, and flying off to collect pollen and nectar than were colonies started by a queen that had mated only once.

A group of scientists at the University of Sydney performed a clever experiment to examine the genetic regulation of reproduction in honeybees. Like humans, bees are affected by carbon dioxide gas, but unlike humans, queen honeybees respond to CO_2 by increasing their ovary development, as if they had just mated and were getting ready to start a colony. In contrast, if a queen is removed from a hive, something the workers can detect immediately, the workers respond to the gas by suppressing their ovary development, just as if the queen were present and producing all the eggs (worker bees are able to lay eggs, although their sisters often prevent them from doing so).

The researchers, led by Graham Thompson, placed virgin queens and queenless workers in a chamber with CO_2 for 10 minutes and then compared the gene expression in the bees' brains as well as their level of ovary development at intervals of several days. They examined twenty-five genes and found differences in expression in ten of them, suggesting that the bees are exquisitely sensitive to small changes in their environment and that the actions of their genes are altered accordingly.

Where did the extreme social behavior in these insects, with its self-sacrificial sterility, come from in the first place? The study of the honeybee genome, as well as detailed information on the genes of social and nonsocial species, supports an idea that had been around for a while among entomologists: start with mothers sticking around to feed their young, and go from there, progressing from maternal care to the more generalized care of siblings. Many insects show a more modest amount of social behavior than the ants or honeybees, as I describe in the chapter on parental care; they may guard their eggs, bring food to the developing young, or join forces with other females to rear offspring collectively, and they provide good test cases for this idea. Toth, Robinson, and a group of colleagues used the common paper wasp to see if care of sisters and care of young were governed by the same genes. Although the genome for the wasps has not yet been sequenced, the scientists used an innovative technique to characterize short segments of DNA that were already known to be associated with social behavior in honeybees. Although the bees and wasps last shared a common ancestor 100 to 150 million years ago, the genetic material that was examined turns out to be amazingly unchanged.

Paper wasps do not show the extreme differences among castes seen in ants or honeybees, but the scientists were able to examine DNA from four kinds of individuals. Foundresses are the females that start up a colony in the spring, usually by themselves, which means they forage as well as reproduce. They rear the first generation of daughters, who then become workers, allowing the foundress to become a queen and spend all of her time laying eggs. Finally, gynes are females that mate late in the year, spend the winter in a sheltered place, and then emerge in the spring to become foundresses.

Despite the identical outward appearance of the four types, and the fact that in some cases they are actually the same individuals performing different tasks, the researchers found that the wasp females differed markedly in the expression of genes in their brains. Workers had brains that were more like the foundresses that also cared for young than the queens and gynes that reproduce. Some of the genes that differed in expression were related to the production of insulin, an important component of nutrient regulation in insects, as in humans, which suggests that becoming social involved evolutionary changes in how food is perceived and processed. Toth and Robinson believe that the path from completely solitary to intensely social made use of a kind of molecular toolkit common to the ancestors of both kinds of behavior, modified in small ways as natural selection acted on the components. This differs from earlier ideas that new behaviors needed new genes.

The Collaborative Dictator

RESULTS such as these are leading us to a much better understanding of what it means to have genes control anything, whether that is social behavior or eye color. People often assume the existence of a gene "for" a trait, so that if you have the monogamy gene, for example, you won't cheat on your spouse, but if you lack it, your infidelity is inevitable. Studying genomes shows this is futile. First, genetic material is often redundant, nonfunctional, or just plain disassociated from any obviously useful protein. Second, genes are the great recyclers — all of our genes were modified from preexisting ones, with some new mutations that occurred at random thrown in. The genes associated with parental behavior are related to those that make a bee more likely to feed her sister, which are also associated with myriad other behaviors. This means that no gene can be associated de novo with a single trait and that trait only. Third, and maybe most important, genes are regulated with a complexity that is only just beginning to be understood. As in the paper wasps, it's not the genes themselves that change, it's the conditions under which they are expressed, and that regulation requires a host of other genes.

This is not to suggest that we shouldn't try to explore the genetic basis for behaviors such as courtship or maternal care. On the contrary, the new technologically sophisticated methods can reveal extraordinary detail about the mechanisms behind even complicated behaviors. But we should abandon, once and for all, the antiquated notion that we will ever have a catalogue of genes that can be neatly assigned to one and only one characteristic, that a gene associated with long eyelashes will have no truck with one making us more likely to prefer salty foods. Genes may dictate the production of proteins, but they do so in a maze of collaboration with other bits and particles of DNA.

What Next?

GENOME sequencing seems to induce a kind of greed in scientists, a hankering for more species with more variants of behavior and appearance. Many biologists have a favorite study organism and so often would love to have "their" animal or plant sequenced next. As the costs of processing samples decreases, the need to set priorities won't be quite so pressing, but right now several scientists have come up with justifications for "wish lists" to help guide future efforts.

Evans and Gundersen-Rindal used four criteria to evaluate groups of insects for their place on the list. First was genome size: smaller genomes are easier to sequence, and we already have an idea of genome size for many of the major categories. As mentioned earlier, flies, butterflies, and the bees and ants all have relatively small genomes, while grasshoppers and crickets, cockroaches, and silverfish, those odd little wingless pests in libraries, all have rather large ones. The central database called GenBank already has information on proteins in some of the groups, particularly flies, which also helps in starting a sequencing project. Evans and Gundersen-Rindal also ranked the insects for species diversity within each group, arguing that we would be better off working with diverse groups because they are likely to have more researchers working on them. Finally, they scored the insects for their effect on humans, where, as you might imagine, the elusive silverfish were pretty low on the scale. Overall, they plumped for more flies, more social insects such as bees and ants, and more beetles, with some moths and butterflies thrown in as well.

Beetles, particularly dung beetles, were also favorites of biologists Ronald Jenner and Matthew Wills, who suggested that the horned dung beetles in the genus *Onthophagus* would be particularly useful. As with antlers on deer and moose, the horns are more developed on males and are used in fights between rivals for females, allowing researchers to examine the genetic control of sexual differences. What's more, horn size is influenced by the environment in which a beetle matures, with better nourishment yielding more impressive weaponry; this could yield insights into the ways that genes are switched on and off by external factors.

Using criteria roughly similar to those of Evans and Gundersen-Rindal, myrmecologist and insect photographer Alex Wild mused about which ants would make the best candidates for genome projects. He settled on seven prospects, including the leafcutter ants of the New World tropics, which as their name implies slice off bits of vegetation that they bear off to the nest, where the material is chewed and used as a base for fungus gardens. The wood ants were another favorite, with many examples of social parasitism, potentially giving insight into the evolution of this unusual life history. One of the responses to Wild concurred with his proposal of another species, the bullet ant, which has an exceptionally painful sting, although the enthusiasm seemed to stem more from a desire for revenge by a victim than any biological justification.

The future clearly contains no shortage of animals to examine. I have a sneaking interest in those silverfish, though. Turns out they have some pretty bizarre mating tactics; the male spins a thread between a vertical object, such as a twig, and the ground and places a sperm packet beneath the thread. He then coaxes a female to walk under the silk, where she picks up the packet with her genital opening. After the sperm have drained into her body, she detaches the packet and eats it. The genetic story behind this kind of sex at a distance must be pretty amazing.

CHAPTER 3

The Inner Lives of Wasps

Personality in Insects

In *The Sword in the Stone,* T. H. White describes how the magician Merlin educates young King Arthur by turning the boy into various beasts: a fish, a snake, a badger. He transforms young Wart, as he is called, into an insect only once, and that only because Wart is confined to his bedroom, with Merlin shouting at him through the door, and more substantial spells have a hard time entering the keyhole. (The logic of this constraint has puzzled me ever since I first read the story as a child, given Merlin's other superior abilities, but perhaps White felt it necessary to find some justification for why one would ever become an insect in the first place.) In any event, Wart becomes an ant, and it is not a happy transformation. Instead of being thrilled at, say, his ability to lift objects heavier than he is, or his exquisite sense of smell, or his remarkable ability to walk on vertical surfaces, the ant-boy is horrified by the lack of individuality among his nest mates. Each ant (with its "mute, menacing helmet of a face") is like every other, obeying the rules of the queen without questions, and a sign above the tunnel states, "Everything not forbidden is compulsory," a slogan that Wart "read with dislike, though he did not understand its meaning."

Although the ants turn out to have many unpleasant qualities, the most chilling one is that they are automatons, with no independent thoughts, designated only by numbers and letters and interchangeable in their repetitive tasks of collecting food and burying their dead. This image of insects, particularly the social species such as ants and bees, is behind countless dystopian views of the behavior of extraterrestrials in science fiction, perhaps best personified, if that is the correct term, by the Borg of *Star Trek*. These cyborgs assimilate all other beings that cross their path, intoning, "Resistance is futile." They have a queen, work unceasingly, and most crucially, like the ants in White's book, lack all individual identity, having sacrificed it for the good of the group.

Nothing is more important to us than our uniqueness as individuals, and we point to our different personalities as evidence of our humanity. Conformity within the tribe may be valued differently across cultures, but no one thinks a society in which personality is subsumed by service to the state is desirable, even the most diehard communist. Individuality is something of an excuse for selfishness. And while we may freely admit to our pets being distinct individuals, or be willing to believe that a particular elephant or gorilla might be brave or shy, confident or anxious, the buck stops firmly at the backbone. Invertebrates in general, and insects in particular, are assumed to be milling masses of sameness. Perhaps the idea of them as automatons is part of what we find so terrifying about swarms of locusts or bees: each individual is seen as interchangeable with every other, so that killing one has no effect on the rest of the group. They just press on, relentless zombies in our fields and kitchens.

And yet, as with so many other stereotypes about insects, this one turns out to be wrong. They do have personalities, or versions of them, which leads us to question not only the function of individual differences in animals, but in ourselves. We may take pride in individuality, but what is it for? And if being individuals doesn't set us humans apart, what does?

Waspishness in Wasps and Boldness in Spiders

ALTHOUGH psychologists can argue endlessly over definitions of personality, most of those definitions contain some version of individuals showing consistent differences in how they feel and behave. Someone who is aggressive today will be aggressive tomorrow, and aggression in the boardroom means aggression on the basketball court. We also talk about temperament, the predilections that seem to be present when we are born and that shape the formation of personality later on; a fussy infant may become an anxious adult. Sam Gosling, a psychologist at the University of Texas who studies personality in animals, notes, "In some cases, the word *temperament* appears to be used purely to avoid using the word *personality*, which some animal researchers associate with anthropomorphism."

It is true that having a personality seems to imply that one has emotions, a slippery slope when referring to animals, although many early researchers, including Charles Darwin, had no trouble with making the leap. Darwin wrote an entire book on the topic, titled *The Expression of Emotion in Man and Animals*, and while he concentrated mainly on mammals and whether, for example, baring the teeth in dogs had its counterpart in human sneers, he did not exclude insects: "Many insects stridulate by rubbing together specially modified parts of their hard integuments. This stridulation generally serves as a sexual charm or call; but it is likewise used to express different emotions. Every one who has attended to bees knows that their humming changes when they are angry; and this serves as a warning that there is danger of being stung."

Far be it from me to contradict the founder of evolutionary biology, but I don't see it this way. Just because a beekeeper—or Darwin—can predict that a bee is going to sting doesn't mean said bee is flooded with rage. It just means that the beekeeper is skilled at reading the bee's signals, like a weather forecaster knowing when a storm is coming because of the quality of the wind and clouds. Animist beliefs and poetic license aside, we don't conclude that the storm is angry either.

Like most modern biologists, I think insects have personality but think it is presumptuous, not to mention anthropomorphic, to claim that they have humanlike feelings. It is simply too hard to know what is going on inside another being's mind, even when that being is another human, and it seems safe to say that whatever an ant is feeling, it probably isn't the exact same thing humans feel. We are particularly hampered by the lack of facial mobility in many animals, including insects, which makes them even harder to identify with. It's hard to look into the eyes of a butterfly and feel a connection with the being within. Reading expressions is key to our assessment of mood and, hence, personal characteristics, so the absence of frowns and narrowing eyelids in many species (not to mention the eyelids themselves) means that we must use other cues in assessing animal personality.

But as I do with so many other aspects of insect life, I find the absence of humanlike emotions both challenging and soothing. Challenging because if insects lack feelings, where do their personalities come from? Insects make us ask more and more exacting questions about them, keeping us from sloppy generalizing that assumes they are just like us. And soothing because the insects exist, complete and lovely, in their own world that works just fine without the rules and assumptions that govern human behavior. We think—we know—we would be horrific shadows of ourselves without emotions, and we assume that our personalities stem from the way we feel. But if insects can have personalities without emotions, we have to look harder for the source of those characteristic differences among individuals. Maybe personalities are just collections of traits, like body shape; one is viewed as heavyset or sylphlike because of the shape of one's limbs, the size of the joints, length of the fingers all working together to give an impression of bulk or slenderness. The personality has different components than the body, but it is all still the same thing, and insects show us a reduction down to essentials once again.

Not all scientists have been willing to accept that insects lack emotions. Donald Griffin, who discovered the details of echolocation in bats and studied animal communication for much of the last half of the twentieth century, spent the later years of his life trying to convince his fellow scientists that nonhumans—whether dolphins, chimps, or honeybees—can possess what he unabashedly called consciousness. He challenged the prevailing view that consciousness, and by extension true emotion, is so subjective that one can never know if others share it. Instead, he defined consciousness pragmatically, as the "versatile adaptability of behavior to changing circumstances and challenges." By that token, insects certainly qualify, and Griffin was fascinated by the complex communication of social insects such as ants and bees. He dismissed concerns that the small brains and differently organized nervous systems of insects precluded consciousness as immaterial, asking,

"What underlies this dogma that only a vertebrate central nervous system is capable of organizing thoughts? . . . The behavior of some insects is far more flexible and versatile than previously recognized. Perhaps this new behavioral evidence will modify our long-standing conviction that all invertebrates are thoughtless automata."

I like the idea that we have been underestimating insects, but I think we are on very shaky ground extrapolating our own feelings to beings so different from us. Calling insects conscious, or saying that their variable personalities mean they are much like people runs the risk of not paying attention to what they are actually doing, and instead assuming that they are little people in chitin suits. In the long run, I find it more rewarding to see them as insects, and leave the question of their awareness alone.

Scientists therefore simply rely on the outward behavior of an animal, often under controlled experimental circumstances, to tell them something about its personality. If you place a mouse in the middle of a bare room, is it likely to explore, or huddle in a corner? Does another mouse do the same thing? And if one animal is an exploratory type, does that mean it is also likely to be exceptionally aggressive toward other members of its species? It turns out that it does. The "bold-shy continuum," with some individuals eager to explore and others more risk averse, has been documented in several kinds of animals as well as humans, with evolutionary biologist David Sloan Wilson providing some of the landmark research on the topic. He points out that it is important to remember that being shy, whether in people or sunfish, his favorite study system, is not equivalent to being a loser; in other words, there is something more going on here than just dominance over food or nest sites. Instead, animals take their place on the continuum independent of other determinants of dominance, for example, how large they are and, hence, how likely to win a fight.

Over the last few years, biologists have also noticed that some individual animals, whether they are fish, ferrets, or fruit flies, tend to show predictable suites of traits, not merely characteristics such as boldness or shyness during a single event. The predictability can happen in two ways. First, an individual that is, say, bold when faced with a predator will also be likely to be aggressive and attack another member of its species, so that its behavior under one set of circumstances predicts a different kind of behavior under another. Second, an animal that is bold today will be bold tomorrow, and one that hangs back will hang back all the time. In another effort to avoid anthropomorphism, or perhaps just because of a fondness for jargon, scientists often refer to these repeatable clusters of traits as behavioral syndromes, a phrase that evokes a bit of the pathological to me — are there behavioral syndrome support groups? Regardless, biologists do use the terms *boldness* and *shyness* to refer to animals, particularly fish, for some reason, and bold sunfish are those that are more likely to inspect a predator introduced into their tank, as well as acclimate more quickly to being in the lab and forage more voraciously. These behavioral characteristics can have long-ranging consequences; the bold and shy fish even differ in the parasites they harbor, probably because the different activity levels mean that each type hangs out in a slightly different environment and is exposed to different diseases.

Be that as it may, scientists don't try to replicate the five commonly used axes for personality in humans: extraversion/introversion, antagonism/agreeableness, conscientiousness, neuroticism, and openness to experience. Too many of those criteria require self-reporting, but equivalents can often be measured, for example, the frequency of fighting that occurs when animals are in a group, or the length of time it takes an animal to move through its enclosure. Gosling cautions that it is important to know the standards against which the measures are being compared; after all, he points out, if one asked whether a deadly black mamba in a room was aggressive, it would be possible to respond that it was not, if "it has attacked only two people in the last hour, well below the norm for this species of snake." Nevertheless, one would still be ill-advised to enter. There is also something called The Horse Personality Questionnaire, with categories of Dominance, Anxiousness, Excitability, Protection, Sociability, and Inquisitiveness; why horses require six descriptors while humans only need five is an interesting question.

Keeping this in mind, a wide range of animals, including several insects and spiders, show consistency in their behavior. For example, fishing spiders live at the edge of ponds and feed on insects near the water. Different individuals attack their prey with differing degrees of alacrity, and the individuals that leap upon their prey more quickly get more food. This seems like an all-around good thing, until you learn that predatory eagerness in females is also associated with a greater likelihood of killing and eating one's potential mate. Fishing spiders also respond to the threat of attack by a potential predator by quickly diving under the water and remaining submerged in an air bubble until the danger is perceived to have passed, a period that can exceed 90 minutes. J. Chadwick Johnson and Andy Sih at the University of California, Davis, found that the length of time female spiders spent submerged varied among individuals, and bold spiders—ones that emerged relatively quickly from their underwater shelter—were also likely to go after food more decisively and to respond to males that were courting them by tapping the water surface.

Because I study them, I am biased toward crickets and have always thought they had plenty of personality, with more of it, not to mention charm, than their relatives the grasshoppers. Male crickets are rather pugnacious, and the ancient Chinese often pitted them against each other in specially constructed arenas, much like miniature cockfights. Some individuals were highly prized as winners, with poems written about their prowess. And indeed, my intuition was upheld; recently, Raine Kortet and Ann Hedrick found that fighting ability in a North American cricket was not only variable among different males, but winners were more brazen about emerging from a refuge in their container after they had been disturbed in a manner simulating a predator.

Water striders are the leggy insects that dimple the water of streams and ponds all over North America; they skate over the surface, grabbing both prey and, during mating season, each other. More accurately, male water striders jump on females and attempt to mate, while the females often try to shake them off. Although they all look similarly jittery to the casual observer, the striders, too, vary among themselves in their level of activity, with relatively sluggish individuals and more perky ones. More active individuals also tend to be more aggressive. Andy Sih, working with Jason Watters this time, created groups of male water striders in seminatural streams by putting like-minded, or at least like-behaving, individuals together, so that some groups had members that were more laid-back overall, and others those that were more likely to hustle. The scientists then put females into the mix and measured the success of the males in obtaining mates. Somewhat to their surprise, the groups that measured the highest on the hard-driving scale didn't end up with the most overall mates. Watters and Sih discovered that such groups were likely to have "hyperaggressive" individuals whose overenthusiastic pursuit apparently drove the females away. As with humans, it's easy to overdo the hard sell.

Even those poster insects for uniformity, tent caterpillars, turn out to have some inner uniqueness. Tent caterpillars live in rather messy webs spun in tree branches and can number in the thousands during outbreaks, when they are serious forest pests. Their munching, marching armies can defoliate tens of thousands of acres. Understanding variations in their behavior is important for controlling them, so the topic has received considerable study, and it turns out that individual caterpillars show consistent distinctive patterns of sluggishness or activity over several days. Admittedly, how much a caterpillar walks or eats during an hour-long observation period, and whether it is more or less than the amount attributable to another caterpillar, is not what most people think of when they imagine an animal with a characteristic personality, but it still differs from that Borg-like image that is traditionally held.

As for Wart's—and T. H. White's—stereotypes about ants, they too may not be well founded. In his delightfully titled 1928 paper *Psychological Experiments with Ants,* G. Kolozsvary studied the escape behavior of the insects and found that they varied in what he termed *nervousness.* Other more recent papers have found individual differences in how ants cared for the pupae in the nest and how they responded to the others in the colony.

So personality is everywhere, even if Arthur, the once and future king, remained unconvinced. One of the most interesting implications of this realization is that scientists are starting to have what might be termed a more holistic view of animal behavior. If how an individual behaves now can be predicted based on the way it behaved before, we should probably stop acting as if every day, and every experiment, is a world made new—looking at an ant or fish or cricket under one set of experimental circumstances isn't independent of looking at that same animal under another set. This means that even biologists should see the animals they study as unique individuals at least sometimes, rather than interchangeable subjects. We've shied away from this before lest we appear anthropomorphic, but now it seems as though there are solid scientific reasons not to assume that all ants are the same.

She Must Get That from Your Side of the Family

WHERE do personalities come from? In other words, are we — and other animals — born with them, or are they shaped by our experiences? It's particularly instructive to ask these questions about insects, because with humans and other cognitively complex vertebrates, it's virtually impossible to disentangle the two. We humans are interacting with others nearly nonstop from the moment we are born, and maybe even before that if the exhortations about talking, reading, or playing music to the developing fetus are to be believed. Other social animals such as dogs or primates are almost the same. But insects have a much more modest amount of input from others, and as I have said repeatedly, we can manipulate their environments much more easily as well. Therefore, any behavior that persists despite a change in juvenile milieu must be genetic, and conversely, a behavior that is different in genetically similar individuals, for example, siblings, that are reared apart is likely to be due to learning.

Like many traits, shyness and boldness seem to be at least partly heritable; if an individual's father or mother was bold, chances are that it will be too, even if it is reared apart from its parents. Using fruit flies, Marla Sokolowski has been able to find not only a single gene, but its coded protein that lies behind a tendency to either move around as a larval fly (*rovers*) or have a more couch potato–like persona (called, reasonably enough, a *sitter*). Humans have also been able to domesticate breeds of horses, dogs, and other animals that possess not only particular body types but also characteristic behavioral traits, including willingness to fetch (retrievers) and aggressiveness (think pit bull), which means that such characteristics must be able to be passed from parents to offspring. Bold and shy people have different responses in their brains when they are presented with the same photographs of familiar or unfamiliar people, suggesting that these differences are an integral part of our makeup.

But as is also the case for many traits, the environment affects how much boldness or shyness (or any other aspect of personality) is expressed. Early experiences such as how much a mother interacts with her offspring can modify the tendency for an animal or person to be reckless or reserved, docile or rebellious. For many insects that lack any parental care, later behavior can still be affected by the place where a mother lays her eggs. In Wilson's sunfish, boldness means that you explore a new object in your pond sooner than the other fish do, and you are more willing to come out of hiding when a predator approaches. These differences between bold and shy individuals persisted in the wild for as long as Wilson and his crew were willing to look for them. And immediately after they were brought into aquaria in the lab, the bold fish were more willing to eat fish flakes, a novel food, instead of sulking in a corner, dreaming of scrumptious snails. But after a few weeks of getting used to the glass and plastic of their new digs, the distinction between the two types disappeared, and the formerly bold and shy sunfish were equally likely to approach a new object. The real world, it seems, keeps us — or at least some animals — different. These results make it tempting to speculate about the homogenizing effects of institutions such as prisons, or maybe even just urban living, on us humans; but of course we don't have a similar controlled experiment on people to use for comparison.

This universality of personalities across many different kinds of animals, including insects with their tiny brains, has two crucial implications. First, it means that the mechanism behind personalities can't be all that important, or at least that different parts of our physiology must account for the existence of personalities in different groups. In people and other mammals, we attribute, maybe even excuse, being laid-back or anxious to our hormones. Our stress levels are up because of cortisol or adrenaline, our neighbor is phlegmatic because his testosterone has decreased as he's aged. It isn't that hormones cause us to have particular traits, but that we have to have some physiological manifestation of our psychic differences. Even things that seem to be all in our head have to come from somewhere in the body, whether that is hormones coursing in the blood or electrical signals leaping in the brain.

Invertebrates, however, lack the same kind of hormonal system that mammals have, so their tendency to dart across a pond or cower under a leaf must arise from a different physiological source. The hormones are a handy means to an end, but they are not the only one. If fish, ants, and crickets have personalities too, we have to look somewhere other than our vertebrate types of tissues and organs for where they come from.

Second, when something seems to have evolved independently over and over again, such as wings in bats, birds, and butterflies, you have to suspect that nature is onto a good thing. So the existence of something that looks so much like our personalities, but is obviously evolutionarily independent, suggests that having a personality serves an important function.

The Face Is Familiar, but the Sting Is Different

IMPLICIT in the existence of different personality types is the ability to tell each other apart. If you can't keep track of whom you're dealing with, knowing that someone is rude and someone else magnanimous is useless. Young Wart was taken aback by the sameness of the ants he encountered, each distinguishable only by a set of letters and numbers (the Borg, too, were designated by numbers, with the voluptuous Seven of Nine, taken into the *Voyager* starship community, using her lack of a humanoid name as an indication of her loss of humanity). The implication was that since all the individuals behaved the same, there was no point in distinguishing among them, either by appearance or appellation. And yet if animals have personalities, and others react to them as individuals, they must not be indistinguishable. We can tell our pets apart, and we are perfectly happy to watch nature shows that differentiate — and name — individual elephants or meerkats. But what about insects?

One roach or ant may look just like the next to us, but when Liz Tibbetts of the University of Michigan puts up a slide of the paper wasps she studies, she calls the array of face shots *portraits,* and she does it unselfconsciously. To her, they are just as distinctive as a series of family holiday photos, or a row of oil paintings of ancestors on the manor wall. And indeed, once you scrutinize the lineup, the yellow triangular faces do differ: a couple of black dots across the forehead of one, a big dark triangle across the chin of another. For over a century, conventional entomological wisdom held that given the large numbers of individuals in a social insect colony, the most one could hope for in terms of individual recognition would be a rough ability to classify other wasps (or bees, or ants) into categories: male versus female, or nest mate, to be fed or at least tolerated, versus foreigner, to be attacked. Maybe it is the opposite of anthropomorphism: instead of assuming animals are like us, we assume they are not. Both are risky generalizations that turn out not to be borne out by the facts. Increasingly, biologists such as Tibbetts are discovering that at least some insects can do far more than peer nearsightedly at their neighbor and call it friend or foe.

Paper wasps, unlike honeybees, live in relatively small groups of females, all of which are capable of laying eggs, so they lack the clear distinction between worker and queen. The females of one of the species Tibbetts studies, *Polistes fuscatus,* fight vigorously for dominance at the beginning of the season. The rank an individual attains is crucial, because the higher up a wasp is in the hierarchy, the more food she can garner, the less work she does, and, most important, the more eggs she contributes to the colony's reproduction, giving her a larger share of genes in the next generation. But Tibbetts noticed that, as with other species that live in socially stratified groups, for example, baboons, the overall amount of aggression in a wasp nest subsides with time, and the wasps do not have to fight each time they meet to reestablish who's boss.

Tibbetts suspected that the wasps used their variable facial patterns to recognize and remember individuals, and she tested her idea in an ingenious if simple way: she painted the faces of wasps so that they no longer seemed familiar to their nest mates. To minimize her risk of getting stung, she nabbed the wasps early in the morning, when the wasps were chilly and less inclined to object to being handled. Once the redecorated insects were returned to their nests, Tibbetts watched the reaction of the rest of the colony. As she had predicted, the wasps were much more aggressive to the altered individuals than they had been before, although the aggression subsided after about half an hour, indicating that they had learned the new facial pattern of their nest mate. It was clear that the wasps were not simply reclassifying the painted females into rough categories, such as "familiar" or "unfamiliar," because paper wasps use chemicals on their outer surface for that purpose; furthermore, a female perceived to be from outside the colony would have been ripped to shreds or chased away, not simply pushed around a bit more than before. What is more, the wasps didn't seem to care how the apparent interlopers were painted — a bit more yellow on the chin garnered no more or fewer attacks than a larger brown stripe between the eyes. That means that the wasps are not using the face patterns as an indicator of size or age or some other quality of an individual, but instead as genuine identity tags; having two black spots, for example, doesn't say, "Stay away from me, I am large and fierce," but rather, "Sam I am" (or Samantha, in this case).

In another species of paper wasp, *P. dominulus,* facial patterns do indicate both size and dominance, and females pay particular attention to the blotchiness of the black marks. A more broken pattern means a robust, higher-quality individual, for reasons that aren't yet clear. This time, Tibbetts painted the faces of the females so as to either make them look more or less dominant, and then allowed the wasps to guard a sugar cube in the lab; in the wild, the wasps eat nectar, but they happily consume sugar in captivity. Pairs of wasps, one painted to look like a subordinate and one to look like a dominant individual, were each assigned their own sugar cube, and then another wasp was introduced to the container. The wasps share food, but dominant individuals are harder to coax into contributing than submissive ones. As you might expect, the supplicating wasp was more likely to choose a cube being defended by a wasp whose facial pattern suggested she was a loser. What's more, when Tibbetts staged encounters between wasps that were strangers to each other, a subordinate individual painted to look like a dominant one was much more likely to be beaten up by the real dominant wasp, showing that even wasps dislike a cheater. Interestingly, other researchers studying the same species of wasp found that body size, rather than pattern, was important in determining social rank in a population from Italy (Tibbetts works in North America); the reasons for the difference are not well understood but may have something to do with the timing of food shortages and, hence, growth of the young insects.

It turns out that individual recognition is more likely to evolve in wasp species that show more complex social behavior than in those with more short-lived or simple interactions. In some species of paper wasps, a single female always starts a nest, and her daughters then contribute to the growth of the colony. In others, several females build the cells and attempt to lay their eggs simultaneously, giving rise to the jockeying for social status described above. Among yet a third group, either approach is seen. Each of these scenarios calls for an increasingly astute approach to social politics. As one might expect, species that are likely to have complicated group dynamics, as indicated by the likelihood that multiple individuals will have to work out their hierarchy, are more likely to show a lot of variability in their markings.

What about Wart's ants? Contrary to his experiences, they too can recognize individuals, although unlike the wasps, they lack the distinctive facial or other body patterns used to tell one six-legged companion from another. Instead, ants use chemical signatures, individual odors that ants produce on their external skeletons. We have known for a long time that ants as well as many other insects use such chemicals as general news bulletins — "I am one of you; let me through," or "Female here, sexually available till six" — but had always assumed that such rough categories of signals were the limit of their abilities. But Patrizia D'Ettorre from the University of Copenhagen in Denmark and her colleagues wondered whether ant queens in one species, at least, might do something similar to what the paper wasps do.

Most ant colonies have a single queen, who settles into a life of fecund bliss after a single mating flight in which she left her home nest, mated, and then spurned any further gallivanting by chewing her own wings off and digging a chamber for her eggs underground. But in a tropical ant species called *Pachycondyla villosa*, which lives in forests from southern Texas to Argentina, several of the young queens (older literature actually refers to them as princesses, though this terminology is not seen much any more) band together to make a nest in a bit of rotting wood and form a society with a dominance hierarchy and division of labor among the members. As with the wasps, it stands to reason that sorting out whose turn it is to take out the trash would be easier if the ants could tell each other apart. D'Ettorre tested this idea by collecting some of these new queens in Brazil and allowing pairs of them to establish a dominance relationship in the lab. The ants fight by biting, stinging, and boxing with their antennae until one party backs off. Then each queen was allowed to interact with either her former companion or an unfamiliar female. In some trials, D'Ettorre anesthetized the ant her subjects were given to ensure that it was really an individual odor cue, and not just generally aggressive or submissive behavior, that could be used to distinguish one ant from another. In all cases, both the dominant and subordinate queens recognized their former partners. Chemical analysis of the ants' exoskeletons showed no relationship between any particular compound and whether a queen was dominant or subordinate, confirming that the ants were not simply reacting to a generalized "alpha ant" smell. What was even more astonishing was the ants' memory: even after 24 hours of separation, a hefty interval for an animal that lives just a few weeks or months, the queens remembered their previous encounters and behaved accordingly when the two females met (whether there was any frantic internal searching for identity, a kind of ant version of "You smell so familiar but I just can't remember who won when we were together," wasn't discussed).

Finally, what does being able to distinguish a tiny blob of black or yellow, and remember who has which variant, mean about the brain of the insects that exhibit such a sophisticated ability? In many animals, the part of the brain controlling a much-used behavior is comparatively larger than in species that seem to need it less; thus, for example, bats and owls have disproportionately large portions of their brain devoted to hearing. Neurobiologists Wulfia

Gronenberg and Lesley Ash checked out Tibbetts's wasps, as well as other closely related species, and found that being able to recognize faces didn't mean having a larger brain or larger visual centers. Interestingly, the wasps with the recognition abilities had smaller olfactory centers. Another part of the brain with the rather peculiar name of mushroom body was larger than expected, but the difference was quantitative rather than qualitative, suggesting that such a capability is nothing that unusual among wasps, and that similar abilities may be discovered if we simply look for them.

Personalities and Evolution

ALTHOUGH individual variation in animal personalities is a somewhat novel idea for biologists, variation itself is not, being the stuff that underlies evolution. Natural selection acts by some variants reproducing better than others, leading to a preponderance of the fitter genes in the population. But this seems like a paradox in the consideration of personality as consistent differences in behavior: if individuals are different, don't some of them perform better than others? And if so, why haven't the more successful ones come to outnumber the less successful ones, so that we are left with only those personality types that are optimal for their environment?

This question is part of a much larger one, namely, what maintains all the tremendous variability that we see in nature. It's all very well and good to natter on about snowflakes, but living things are far more distinctive. Mutations, changes in DNA that occur spontaneously or as a result of environmental forces such as radiation, supply the raw material for evolution to act upon, of course. But mutations are just the original source of variation; differences among individuals persist in populations over many generations.

Scientists have a number of theories about how genetic variation is maintained in populations, ranging from the simple notion that selection may not have had a chance to weed out loser genes in some cases, to elaborate ideas about interactions among the genes that happen to be linked together on the same chromosome. Many of these explanations apply to the maintenance of different personalities, too, particularly the idea that advantages and disadvantages trade off against each other. Take the spider femme fatales, for example. Being exceptionally eager to snag a passing fly means that you are more likely to survive another day, which is obviously favored by natural selection: you get moved up a notch on the "likely to live long enough to have babies" scale. But being too rapacious in your treatment of a potential mate might mean lower likelihood of getting enough sperm to fertilize your eggs, which tips the balance in the other direction. On the other hand, a more suave approach to mates may be irretrievably associated with a more lackadaisical feeding style. Each personality type has its pluses and minuses under different circumstances, and they can all coexist because they each make trade-offs in different ways.

Of course, having a particular set of personality characteristics doesn't necessarily mean giving something up. Some individuals are just all-around winners, and if a behavior is advantageous all the time, then the individual exhibiting it makes out like a bandit and never pays the price. Being particularly active, for example, might mean that an animal finds more food, is more likely to be chosen as a mate, and is first in line for shelter when a storm threatens. Although our Puritanical sides may argue that everything has its price, sometimes that price is negligible. In cases where a trade-off within an individual doesn't exist, one might expect that selection would favor those lucky few, and variation might indeed decrease in the population.

Alternatively, having a particular personality type may be advantageous if your environment is not very predictable. Andy Sih likens it to investing in the stock market; if you don't have any information on what is going to happen, it may be better to keep your money where it is than try to play the market, a particularly prescient point in today's economy. For an animal, being predator-wary all the time can be a good strategy when the actual likelihood of being eaten is unknown, a kind of better-nervous-than-sorry attitude. If on the other hand one can be reasonably sure that predators aren't lurking, then one can let oneself go a bit more, as it were, and act one way under some circumstances, and another way under others.

Yet another way that behavioral variation can be maintained is by having the success of some types depend on how common other types are in a population. If a group of, say, water striders is mainly composed of lethargic, cautious individuals, then being a fast-moving brave strider will be advantageous as long as the reckless types are rare, because the rare ones can dart in and grab food while everyone else is just gathering their thoughts. Once the fast movers reproduce and rise in abundance, however, the slower bugs can in turn do well, perhaps because their type is less likely to be nabbed by a predatory spider. Then the slow individuals do better, and so on; both types can persist over time, even if their relative numbers fluctuate.

Similarly, several researchers, most notably Judy Stamps at the University of California at Davis, suggest that personality tradeoffs can have repercussions for how fast an animal grows and how many offspring it has. If you live hard and die young, but grow quickly and have your children early, you may end up leaving as many genes in the next generation as someone who grows more slowly, dies at an older age, and paced his or her reproduction more prudently over time. In other words, you could have a personality and life history more like a rabbit or one more like an elephant, and still be a member of the same species — an individual with characteristics at one of the opposite ends of the spectrum. Again, both styles can be maintained via evolution. The question is whether having an aggressive or bold personality is linked with a tendency to die sooner or have more offspring, something that bears investigation.

Another implication of personalities for evolution is that they make behavior less flexible. In an article summarizing recent work on animal personalities, Alison Bell from the University of Illinois said, "Animals do not always change their behavior as much as they should." While that "should" seems a wee bit judgmental to me (who are we scientists to decree endless flexibility among our animal subjects?), remember Ralph Waldo Emerson's admonition that "a foolish consistency is the hobgoblin of little minds." It is nice to think that insects and we have an equal opportunity to show small-mindedness, but what is the cause of this limitation? Bell speculates that it might just be too hard to change one's personality under different circumstances, even if it would theoretically be advantageous to do so; shifting around the hormones and nervous system reactions might require too much retooling of basic physiology. If the same evolutionary costs and benefits apply to human personalities, we may be better able to understand why some apparently counterproductive ways of coping in humans persist, and ultimately, why we have personalities at all.

Can We Have "A Feeling for the Organism" if the Organism Lacks Them?

In Evelyn Fox Keller's 1983 biography of Nobel Prize winner Barbara McClintock, some of the geneticist's success is attributed to her "feeling for the organism," a way of understanding her scientific subjects that transcended traditional scientific data collection. Keller describes how the scientist had to immerse herself in the minute details of her study subject, understand "how it grows, understand its parts, understand when something is going wrong with it." The book, titled *A Feeling for the Organism*, became something of a symbol of science needing something more than facts and charts, requiring in addition an emotional investment from the investigator that verges on the mystical.

This all sounds fine and inspirational, until you discover that the subjects of McClintock's fascination are corn plants. Dotty gardeners cooing to their tomatoes aside, identification with anything other than pets and perhaps a few other animals such as wolves or ravens seems like so much New Age claptrap. But according to Keller, McClintock felt that people had a "tendency to underestimate the flexibility of living organisms." If that underestimation is true for corn, it is doubly so for insects. The idea that they are all alike, with identical reactions and identical lives, used to be unquestioned, as T. H. White illustrated. Now, though, we know better. The famous entomologist Vincent Dethier, author of the masterful book *To Know a Fly* and discoverer of many important aspects of insect neurobiology, fretted about the likelihood that his subjects had internal lives. In a 1964 essay about the continuum between insect and vertebrate brains, he said, "Perhaps these insects are little machines in a deep sleep, but looking at their rigidly armored bodies, their staring eyes, and their mute performances, one cannot help at times wondering if there is anyone inside." Nearly half a century later, and with the appropriate caveats about personality sans expression, the answer seems to be yes.

CHAPTER 4

Seinfeld and the Queen

IN *Bee Movie,* Jerry Seinfeld plays a slacker honeybee, yearning for a life beyond the tedium of the factory. His portrayal of a social insect as male is as glaring in its inaccuracy as in its ubiquity. No one, including me, expects movies to be faithfully accurate in all their details. But there are errors and errors, poetic license versus jarring ineptitude, bloopers versus downright stupidity. Talking animals is one thing, but getting it wrong about honeybees is on a par with portraying astronauts in a universe where the earth revolves around the sun, or TV doctors who are worried about anemia due to a lack of lead in the blood instead of iron.

And it goes beyond an expert's superiority at having "gotcha," more than the fun of discovering a telephone in a pre-Bell movie. The way I see it, there are at least two problems with exhibiting such flagrant ignorance about the sex of the social insects. The first is that it perpetuates a skewed vision of the world and its sex roles, a vision that can end up doing our own society some harm. The second is that if you assume everything in the insect world is the way it is in our own species, you miss out on stuff. And in sex roles, as with many things about insects, the truth is much, much more interesting than fiction.

His Thighs with Sweetness Laden

NOT KNOWING that virtually all the ants or honeybees that one sees are female is nothing new. The phrase in the subtitle above comes from a poem by Charles Stuart Calverley, a mid-nineteenth-century Englishman said to be the father of the "university school of humor," a designation that as a professor I find compelling yet enigmatic. He wrote several books of poems, including *Fly Leaves,* which contains the following lines:

> When, his thighs with sweetness laden,
> From the meadow comes the bee,

Even Benjamin Franklin fell into the stereotype; a letter he wrote to a woman he was apparently courting contained the following lines from a poem by William Pulteney:

> Belinda, se, from yonder Flow'ers,
> The Bee flies loaded to his cell:
> Can you perceive what he devours?
> Are they impair'd in shew or smell?

Franklin and Calverley, as well as the people from Disney, Pixar, and various other movie studios, were following in a tradition present since at least the ancient Arabs and Greeks, who believed that a king bee, or "bee father," in the case of the former, was in charge of the hive and that the followers were probably male as well, though this latter point was the source of some debate. The Greeks were able to distinguish a category of bees, the drones, that were larger than the workers, but although they disapproved of the bees' apparent laziness (the drones hang around the hive until their mating flight and are fed by the workers but do not collect nectar or pollen), they could not determine the drones' sex. Part of the confusion seems to have been that the Greeks were well aware of the stinging ability of bees and found it impossible to believe that any animal bearing such a weapon could be female.

Similarly, on viewing the fierce defense of the hive exhibited by most social insects, including honeybees, many Arab cultures likened the colony to an army, which naturally implied a military rule, with males as both officers and soldiers. (The stinger is a modified ovipositor, or egg-laying structure, but the workers are generally unable to reproduce.) Aristotle tried to reason out the problem but ran into difficulty because if the stinging bees, the workers, were male, that would have suggested that the drones were female, and he was unable to accept the idea that the males in a society did all the work of taking care of the young. He eventually concluded that bees might have the organs of both sexes in a single individual, as do many plants, but then was further flummoxed by the lack of reproduction on the part of any individuals but the queen (or leader, as he called it).

In *Henry V*, Shakespeare refers to "The lazy yawning drone," as well as the king of the bees, again without any hint of recognition that the shiftless members of the hive were male. Writing in the nineteenth century, Calverley, it turns out, actually should have known better, because the sex of the queen was ultimately determined in the late 1600s. Several writers and beekeepers had guessed at the truth, but Jan Swammerdam, a Dutch microscopist, is generally credited with demonstrating that the individual assumed to be the king had unmistakable ovaries and was responsible for generating the other bees in the colony. Swammerdam published two books with breathtakingly detailed drawings of the anatomy and life cycles of bees and ants, among other things, some of which were not equaled until the twentieth century. He had to manufacture his own miniature tools and adapt the primitive magnifiers of the day for his own use. He is said to have publicly dissected a purported king bee in 1668, an event I must confess I have a hard time imagining. Admittedly, learned men had a different means of spreading knowledge back then, but it is entertaining to picture the way in which news of the impending feat might have spread: "Hey, did you hear? Ol' Jan Swammerdam is cutting open a bee next Tuesday! Who knows what peculiar structures he will reveal! Let's go watch — I'll buy the mead." Do you suppose he sold tickets?

In any event, Swammerdam correctly distinguished the anatomical differences among the larvae, or young bees, and the drone, worker, and queen. The next puzzle was to determine how exactly the queen produced the other types, since no one had ever seen bees engaging in sex. Here he was not so prescient. He suggested that the sperm from the drones somehow wafted through the air and the odor then was powerful enough to impregnate the queen, a theory called *aura seminalis*. Swammerdam noted the distinctive smell of the drones in the colony and assumed that this stench was powerful enough to inseminate at a distance. Why he was content to speculate rather than test his theory is not clear, given his modern, experimental approach to the rest of his work. At one point he even suggested a possible test of the *aura seminalis* idea, in which one would determine "whether the female Bee, enclosed in a little net made of fine thread, or in a small glass vessel covered with a piece of fine linen, or in a box with holes in it, could be impregnated by the bare scent of the male."

In Swammerdam's defense, the mating process in honeybees was not clarified for a few hundred more years, although some eighteenth-century scientists had noticed that a queen bee sometimes returns to her hive with the genitalia of the drone still attached to her reproductive organs. Finally, in the mid-1900s, naturalists discovered drone swarms, groups of virgin male bees that congregate in small areas near hives. At the right time of year, if you know where to look, you can find the would-be suitors by listening for their humming. My college entomology professor took us to one such gathering, and although we could hear but not see the bees, when he flung a pebble into the air above our heads it was instantly pounced upon by the drones. The young queen flies into the swarm and is pursued by the males; as the queen flies faster and farther, she leaves behind all but the most ardent males. Finally she mates with one or more of the drones, who die immediately and fall to the ground while she carries their sperm back to the hive.

Sex and Honey

PARALLELS between bee and human society and gender roles — or lack thereof — have been popular for a long time. The bees' communal life, industry, and apparent self-sacrifice for the good of the colony were held up by many ancient civilizations as models to which humans would do well to aspire, and beehives have served as symbols for groups as disparate as Masons and Mormons. Once it was established that the worker bees did not mate, their chastity was also suggested as an inspiration to women. After the queen bee's sex was discovered, writers such as Charles Butler, author of *The Feminine Monarchie,* published in the early seventeenth century, began to shape their newfound knowledge into a form more acceptable for the social mores of the time. Butler approvingly noted that the drones had a louder voice than the female workers, just as the rooster's crow is louder than the sounds made by hens, but seems to have conveniently ignored the "piping" noise, much more pronounced than the humming of the drones, produced by queens shortly after they emerge as adults from their waxen cells. Piping queens also show another less than feminine trait when they attempt to kill any other queens emerging in the colony at the same time, but this behavior also seems to have gone unremarked. Similarly, queen bees were described by Richard Remnant in his 1637 writings as "gentle and loving," though it is hard to see a female that dispatches rivals by biting their heads off and renders her subjects sterile as being particularly benign.

It is possible that some of the sex role confusion about bees was deliberate; scholars have questioned whether, for example, Benjamin Franklin was unaware of the current state of knowledge about the sex of the worker bee, or whether he might have changed the pronouns from the original song to suit his purposes. I am more inclined to attribute the errors to ignorance. My own admittedly unofficial polls show that, far from being common knowledge, many of my students and those I meet at other universities have no idea that worker bees and ants are female. After my lecture on army ants, the legendary voracious consumers of everything in their path, a typical exchange with a student will go as follows:

"So, Dr. Zuk, you know about the ants?"

"What about them?"

"Like, you said all the workers were female, but what about the army ants?"

"Their workers are female, too."

"Yeah, but what about the soldiers, the ones you said have those huge jaws and everything?"

"They are female, too."

"So really, the soldiers, they are female?"

"Really." At that point the student usually slouches off, eyeing me skeptically. I always feel that I have let them down, but it's not clear how.

Lest you think this is an American fixation, I should tell you about Yamba the Honey Ant. Honey ants, sometimes called honeypot ants, are found in the arid regions of Australia and a few other parts of the world, where they live underground in a network of chambers and tunnels. While most of the ants are able to leave the colony to look for food, some of the workers have immensely swollen abdomens that serve as living storage vessels for the rest of the colony. They never leave their underground den, and other ants tap on the honeypot individuals to get a drop of food. Native peoples, including aboriginal Australians, dig up the nests to harvest the stored honey.

In central Australia, a children's television show features Yamba the Honey Ant, a cheery character portrayed (with a bit of poetic license) in the red, yellow and black colors of the aboriginal flag. On a visit to Alice Springs, I was encouraged to see that Yamba was accurately depicted with six legs, and I wondered if this faithfulness to reality went so far as Yamba's sex. Alas, it does not; Yamba is firmly a male ant, leaving generations of Australian schoolchildren to grow up with the same misconceptions as my own undergraduates.

Insects, Chromosomes, and Surf City

THE FOCUS on males in films about bees and ants, and our own unthinking assumptions about the sex of animals in nature, may say things about the sex bias in human society. But an even more egregious failing in the error is that it means people don't learn about one of the most amazing things in nature: the sex ratio, or the relative proportion of males and females in a population, and how it evolved. Insects have been essential in our understanding of this fundamental issue in biology, and they also exhibit some of the weirdest maneuvers on the basic theme.

We take for granted that under natural circumstances, roughly equal numbers of boys and girls are born into the world. Indeed, for many species of animals, including people, that is the case, with a sex ratio of 50:50. But why should that be so? If you think about it, from an evolutionary standpoint it seems puzzling that so many males are produced, since of course a single male can impregnate a great many females. In species in which males don't have anything to do with the offspring after mating, which includes most animals and certainly the vast majority of insects, it would seem to be much more efficient to produce a few males to provide the sperm and leave it at that, putting all the rest of one's reproduction into daughters, the real producers of the next generation. Why, then, are equal sex ratios so ubiquitous?

If you remember high school biology and a bit about sex chromosomes, you might think the answer is a consequence of how sex is determined when sperm meets egg. Humans and other mammals all possess two sex chromosomes: in females, two copies of an X, and in males, a single copy of the X and a smaller Y chromosome. Since only one of the sex chromosomes goes into each sperm or egg cell, half the sperm make daughters, with two Xs, and half make sons, with one X and one Y. Ipso facto, you get an equal distribution of males and females. Although birds, butterflies, and a few other animals have the situation reversed, with males produced from two Z chromosomes and females from a Z and a W, the same process takes place. The analogy with tossing a coin has been made so often that you have to wonder whether, if we commonly had more than two sexes, we would all be using three-sided coins.

But satisfying though this explanation may seem at first glance, it is ultimately not the answer to the question of equal sex ratios, as Mike Majerus points out in his book *Sex Wars.* First, it's quite possible that the mechanism for determining which chromosome goes with which sex evolved as a means to get to the optimal sex ratio, not beforehand; as with the coins, it would be fallacious to argue that we have the two options of heads and tails on our coins so that we can calculate the sex ratio. Second, as I will describe in more detail below, many animals have the same sex determination mechanisms we humans have but still produce wildly biased sex ratios. And finally, even though "our" XX/XY (or ZZ/ZW) system is widespread, numerous other ways to determine maleness or femaleness are found throughout the animal kingdom, including special genes and the temperature at which eggs are incubated (in many kinds of turtles, for example, cooler nest temperatures yield male babies and warmer ones, female babies). So the equal sex ratios in turtles, at least, must arise from some other source.

As with so many other ideas in biology, Charles Darwin both identified the problem and proposed a plausible solution. Many biologists overlook his original suggestion because it appeared in an early edition of *The Origin of Species* and was retracted in the later ones. Sir R. A. Fisher, a British geneticist and statistician, is generally credited for the breakthrough. Either way, the key is to think about things, not from the standpoint of what is most efficient for the population or species as a whole, but from a gene's eye view. What will make it more likely that a gene will be perpetuated in future generations, the key to evolution?

Say that, indeed, fewer males than females are produced in a hypothetical population of animals, as it was in Jan and Dean's 1963 song "Surf City" ("Two girls for every boy"). Each male fertilizes multiple females, so more copies of each male's genes appear in subsequent generations. That means that having sons is advantageous, so any tendency to do so will be favored by natural selection. Eventually, though, more males than females appear in the population, and then those males are not in nearly as happy a circumstance. Because each child can only have one mother and one father, some of the surplus males must go without mates, and then parents who produce daughters, rather than sons, are at an advantage. That bias then leads to an abundance of females, with the same benefit to being a male that we started with. This kind of seesaw evolutionary process across the generations is called frequency-dependent selection, and it is easy to see that it makes a 50:50 sex ratio the equilibrium that a population settles at, all else being equal. "Surf City," idyllic though it may have sounded to the beach crowds of the 1960s, just isn't a viable way to run a species.

My Sister, Myself

So how do the honeybees get away with having female-dominated societies? The drones comprise less than 5 percent of the total number of bees in the hive, so they clearly violate the equilibrium ratio. In many other insects, although the sex ratio is not always so extreme, a preponderance of females is the rule. The reason the bees can do what we can't has two sources: the way sex is determined, and a special violation of the "all else being equal" clause above.

Unlike the species discussed above, bees, wasps, and ants don't make males and females using combinations of special sex chromosomes. Instead, males are produced from unfertilized eggs, and hence have only one copy of each chromosome, while females have a more normal (to us, anyway) complement of two copies, because the queen produces them by fertilizing eggs with the sperm of the hapless drone or drones she left behind when she entered the hive to start her monarchy. Again unlike humans, queens of social insect colonies can store sperm for years, doling out a son here, a batch of daughters there. The daughters may become workers, in which case they are sterile, with undeveloped ovaries, or future queens, in which case they are cosseted by their sisters and fed exclusively on royal jelly, a nourishing compound (for bees, anyway) that alters hormone levels and facilitates ovary development. (Little or no evidence exists that royal jelly, whether applied topically or eaten, does anything for humans besides diminish their pocketbooks.)

This genetic peculiarity means that in cases where a queen mates with a single drone, her daughters are more related to each other — their sisters — than they are to hypothetical daughters of their own. To understand this, recall that in humans and other species with XX/XY sex determination, offspring get half their genes from their mother and half from their father, making siblings half like each other as well. Because each sperm and egg cell contain only half of each parent's genetic complement, in species such as our own, each sibling has a 50 percent chance of getting any one chromosome from the parent. But because male bees have only one copy to begin with, all the sisters get the same genes from their father. They get the usual 50 percent from their mother, giving them a total of 75 percent of their genes in common. The queen, on the other hand, shares the usual 50 percent of her genes with either her sons or her daughters.

This unusual sisterly closeness is thought to play a role in the extreme altruism exhibited by many of the social insects, but it also is important in the sex ratio. I have been acting as if sex ratio is concerned only with numbers, but in fact, the mere number of males versus females is only part of the story. What really counts is the investment in terms of energy that is made in either sex. Say that females "cost" less to produce — perhaps they are smaller than males, and so fewer calories go into manufacturing them than are required to make the same number of males. Natural selection will favor making more girls than boys, because each girl is cheaper, but the overall investment in each sex is still equal.

For the bees and ants, the asymmetry in how closely related sisters are compared with mothers and offspring means that the evolutionary payoffs for producing different sex ratios are different for the workers than they are for the queen. The workers' only chance at perpetuating their genes is via the future queens and drones, or males, produced by the queen, because workers are sterile. The queen's genetic future lies in the same individuals, but the two differ in what will most benefit each. From the workers' perspective, more of their genes will be passed on if more energy is allocated to the future queens, their sisters, than to their brothers, because they share only 25 percent of their genes with the latter, a third as much as they have in common with their sisters. The queen, in contrast, is equally related to her sons as her daughters, so an equal ratio of investment in the sexes would benefit her the most.

Bob Trivers, an evolutionary biologist at Rutgers University, was interested in determining who "won" this potential conflict between the workers and the queen in the sex ratio of the social insects. With the assistance of Hope Hare, he painstakingly weighed the future reproductive individuals in the nests of a number of ant species. If the queen was in charge, as it were, and her reproductive interests are paramount, one expects that the weights — a reasonable gauge of investment — of the male and female future queens and males would be equal. If the workers prevailed, however, one expects a bigger investment in the young queens, by a factor of three, since the sisters are three times more closely related to each other than to their brothers. It turned out that the combined weight of the males in the nests was just about exactly one third of that of the future queens, supporting the idea that workers control the sex ratio of the colony. That discovery in turn means that insect societies, the most complex social systems on earth, are not dictatorships, but are controlled by a Machiavellian network of alliances and favors exchanged.

Incest and the Solution to Physics Envy

THIS successful application of theory to nature was only one of many triumphs of sex ratio theory as it applies to insects; arguably, the ability to predict with quantitative precision the relative numbers of males and females in an insect group is one of the most impressive confirmations of the operation of natural selection. Ecologists and evolutionary biologists often have to settle for qualitative predictions about the real world: we can say with assurance that a population will grow if more food is available, but exactly how much? Under most natural circumstances, too many other variables are at play to make precise predictions from such a biological hypothesis; the population's growth rate depends not only on food but also on, for example, the likelihood that disease will infect its members, or the abundance of predators in the neighborhood. This uncertainty has led to what people sometimes call "physics envy," the longing for the kind of mathematical confidence seen at least much of the time in the so-called hard sciences.

But physics envy can be cured by a dose of tiny parasitic wasps and an understanding of sex ratio theory. The wasps in question lay their eggs on other living insects, usually caterpillars or fly larvae, and the young wasps then develop inside this live food source until they are ready to become adults themselves. Just after the wasps become sexually mature, they mate, still on their host, and then the fertilized females go off to find victims of their own. Each caterpillar can support only a few females' offspring at most, which means that the young wasps find themselves in a very restricted universe in terms of dating opportunities. It's like going to a singles gathering with just your brothers and sisters and maybe the kids from the house across the street, and knowing you will have to find your lifetime mate from among only those individuals. Although inbreeding, or mating among relatives, has genetic penalties in many organisms, including humans, in the wasps it does not carry the same consequences, and they will mate quite unconcernedly with their siblings.

The mother wasp laying her eggs on the host has the same evolutionary problem as other organisms: how best to leave the most genes in the next generation. If she is the only individual parasitizing a particular caterpillar or maggot, she will get the most bang for her genetic buck by producing only enough males to fertilize their sisters, and making most of her offspring female. Making more males will only produce competition among them for the same mates, which won't do the mother wasp any good. The more additional females that use the same caterpillar, however, the more advantageous producing more sons becomes, because those sons can then compete to fertilize the daughters of those additional females. So one would expect the sex ratio to become increasingly male biased as the competition heats up.

Starting with work published by the evolutionary biologist W. D. Hamilton in 1967, scientists have worked out exactly what sex ratio a female parasitic wasp is expected to produce under a variety of circumstances, such as the number of other females laying eggs in the same host. And nature has been astoundingly obliging in supporting the predictions, down to the last egg. Much of the original research was done in the lab, where the wasps can be induced to lay eggs on fly larvae in dishes, but recent DNA analysis of over three thousand offspring from forty-seven mother wasps collected in the field in Europe confirmed that the equations can be used to predict life in the real world. The theory even applies to the one-celled organisms that cause malaria, which also come in two sexes. Take that, string theory.

A Good Year for Sons

ALTHOUGH on average the number of men and women is roughly equal, the availability of either sex depends on the situation. You will be more likely to find a single man in a bar in Alaska, more apt to meet women at a convention for nurses. As it happens, these concerns about the sex ratio among humans also affected the development of another major theory about sex ratio in nature itself.

It all started when Bob Trivers, the biologist who worked out the ratios of reproductive individuals expected within ant colonies that I discussed above, was a graduate student at Harvard. Trivers was a teaching assistant for a popular course in primate behavior, and as he relates in his collected papers, he had a mathematics student named Dan Willard who took the course as a way to meet women. The math graduate program at Harvard, as elsewhere, was rather like the aforementioned Alaskan bar in terms of its sex ratio (I assume, without really knowing, that the resemblance stops there), but the primate behavior class of nearly three hundred students was about two-thirds female. Trivers never divulges whether Willard's social hopes were fulfilled, but after a lecture about why the sex ratio is usually 50:50, Willard came up with an idea that the two of them later published as a highly influential paper in *Science*.

Like many keen insights, this one seems simple once you hear it. Having offspring is costly, in the sense that it requires energy from the mother to produce eggs or babies. And as pregnant women the world over are acutely aware, the condition of the mother affects the condition of the baby, often for a long time after birth. Even among insects, better-nourished mothers can often produce larger eggs that in turn develop into more robust larvae. Although of course it would be ideal to always be in the best shape possible and produce the highest quality offspring a mother can, the situation doesn't always allow mothers to be in that tip-top condition.

Trivers and Willard reasoned that the consequences of producing a baby that is of less than optimal condition will differ depending on the sex of that baby. Because males in many animal species compete vigorously for mates, only males of the highest quality are expected to be successful in fertilizing a female. Producing a weakling son is therefore unlikely to yield any reproductive payoff for the mother. On the other hand, if one's son is very successful in combat with other males, he can potentially sire many more offspring than a single female, even one in the best condition. Even a daughter in poor condition, however, will almost certainly find a mate and reproduce. So Trivers and Willard predicted that when circumstances keep the mother from pouring a lot of resources into her offspring, she would be more likely to have daughters than sons, and vice versa for mothers lucky enough to be at the top of their game.

The mechanism behind such a sex bias is not entirely clear. No one, least of all Trivers and Willard, suggests that animals alter the usual method of chromosomal sex determination. But many more eggs are fertilized than end up implanting in the uterus or developing into offspring, and it is possible that embryos may be selectively retained or discarded depending on the mother's condition. This does not require a conscious decision on the part of the mother, of course, but selection may have favored implantation of a male or female egg only if, say, her hormone levels reflect a particular level of nutrition.

The Trivers-Willard effect has been demonstrated in a wide range of animals, from deer to fish to birds, and may even operate in humans. A 2008 study of 740 British mothers found that the women were more likely to give birth to a boy if their diets were more nutritious around the time they had conceived. The bias was not huge — 56 percent sons among the mothers that ate the most, versus 46 percent sons among the least-nourished women — but it suggests that more than chance may play a role in the sex ratio even in our Western societies. These insights in humans wouldn't have come about if we hadn't had insects, with their wildly variable lives, as test cases.

Insects do not become pregnant, of course, although some retain the eggs inside or on the mother's body after they are produced. Furthermore, among many insects, large females are favored by natural selection because they can lay more eggs. Good conditions might be expected to cue the production of daughters, rather than sons, to take advantage of the resources necessary to manufacture a robust future mother. Indeed, among many parasitic wasps and flies, the sex ratio is biased toward daughters when the host maggot or caterpillar is large, and toward sons when the host is puny and provides less nourishment for the growing parasites.

Finally, the sex ratio in some insect species can fluctuate wildly, even over the course of a very short time. A Polynesian butterfly had only 1 percent males in 2001, due to an infection by bacteria that selectively kill male embryos. But just 5 years later, researchers found a nearly 50:50 sex ratio on some of the islands where the butterfly occurs, even though the bacteria were still present. Selection apparently acted to reduce the male-killing ability of the bacteria, probably because producing males was enormously advantageous in the highly female-biased populations.

See what you miss if you assume that Jerry Seinfeld makes a good bee?

CHAPTER 5

Sperm and Eggs on Six Legs

Do you suffer from fertilization myopia? Just when you thought you'd heard of all the latest trends in maladies, from attention deficit disorder to cyberchondria (looking up dire diagnoses online at the first sign of a sniffle, in case you didn't know), here comes a new condition to worry about. Luckily, although many of us do, in fact, show signs of fertilization myopia, it can be cured without a single infomercial-shilled medication. All that's needed is a better understanding of insect sex, which might also help us understand sex in other creatures along the way.

Fertilization myopia is a term coined by Bill Eberhard, a biologist who works in Panama and Costa Rica on a wide variety of spiders and insects. For the last twenty years or so, Bill has been intrigued — some might say obsessed — by animal genitalia and the finer details of insect mating. Despite what you might think given this predilection, he is a gracious and genial man and is married with children. He just happens to have an abiding curiosity about the natural world and an unwillingness to accept the conventional wisdom regarding mating behavior.

Until quite recently, that conventional wisdom held that once a male and female mated, from an evolutionary perspective, it was all over. Sperm had been transferred, and now all that remained was to wait for the offspring to appear and carry on their parents' genes. Fertilization was the goal, and we didn't look beyond it. Even in humans, people assumed that the exciting part was the lead-up to sex: the partner choice, the foreplay, the act itself. The aftermath was just an ignominious anticlimax (so to speak) of damp sheets and flaccid organs. Pregnancy may or may not result, but there was nothing anyone could do to influence its likelihood once the deed was done. Arguments about how he wanted to roll over and sleep while she was still wide awake and needing to cuddle notwithstanding, postcoital activity just didn't get a lot of press.

In insects, however, and maybe in many other animals as well, fertilization is far from the end of the mating story. Many insect females, from butterflies to beetles, mate with more than one male in succession before they lay their eggs. This fact had been well known among biologists, but it wasn't until 1970, when Geoff Parker at the University of Liverpool wrote a landmark paper about what he called "sperm competition," that the consequences of such multiple mating began to be fully considered. Parker pointed out that while male competition for females is more commonly associated with the more flamboyant battles between bull elk or elephant seals, it could still occur after copulation has occurred. The males just continue to vie for the prize of siring offspring via the one-celled messengers of themselves they leave as a consequence of mating: their sperm.

The process would, Parker recognized, lead to different kinds of selection on males. On the one hand, male attributes that allowed their sperm to win at fertilization by circumventing the efforts of other males' sperm would be favored by selection; on the other, males that could prevent a female from mating with another male in the first place would do well because they would avoid the whole problem from the start. Insects are ideal candidates in which to observe such postmating activity because the females of most species mate with more than one male, often in rapid succession, and because in many insects females have specialized organs that serve as holding tanks, keeping the sperm in reserve until it is used to fertilize the eggs hours, days, or even weeks later.

The idea of sperm competition appealed to biologists, most of whom, at least at the time of Parker's insight, were male. Numerous mathematical models about the conditions under which a given male's sperm might be favored were developed, and the details of sperm structure in various species—which turn out to vary enormously, as I will explain later—were examined. But other scientists, including Bill Eberhard, pointed out that this emphasis on male competition missed the other half of the equation: the female. After all, it was the female that did the multiple mating that allowed sperm from more than one male to be in the same place at more or less the same time, and it was the female's body in which all the action occurred. Not to mention that the female too has a stake in which male sires her offspring.

So Eberhard and others suggested that females could influence the likelihood that a given male actually fathered her offspring, even after he had done the deed. This biasing of paternity after copulation is called *cryptic female choice*, a term originated by Randy Thornhill at the University of New Mexico. It is cryptic because it takes place out of view, inside the female's reproductive tract. Eberhard went further and pointed out that among insects and spiders at least, we should see that females control much of what happens in reproduction, and that we should stop focusing so short sightedly on that moment when sperm meets egg. In true infomercial fashion, we should wait, because there is more. Much, much more. The musician Björk said, "Football is a fertility festival. Eleven sperm trying to get into the egg. I feel sorry for the goalkeeper." One could, of course, take this the other way and point out that in fertility, both the goalkeeper and the players, not to mention the playing field itself, have a great deal to do with the outcome of the game. It isn't enough to just throw the team onto the field and wait for a goal.

Chemical Genitalia and an Embarrassment of Riches

My good friend and colleague Leigh Simmons claims that you don't understand life unless you have studied dung flies, preferably by actually coming into close contact with the substance that the flies call home. "Buckets of dung," he says cheerfully. "You need to really get your hands in it." Despite the numerous other likes and dislikes we share, I have never been convinced about this one enthusiasm, but I will concede that an understanding of sex in dung flies is crucial to an appreciation of what can happen after sex but before the production of offspring.

As the name suggests, dung flies use cow or other animal droppings as a nursery in which to raise their young, and during summer, the female flies of one well-studied species, the yellow dung fly, seek out the freshly produced pats in meadows all over northern Europe. Once they arrive, they are immediately pounced upon by the males, which have been performing surveillance flights on the dung. As Leigh puts it in his book *Sperm Competition and Its Evolutionary Consequences in the Insects,* "On capturing a female, males will begin to copulate immediately. Struggles for the possession of females are intense. Searching males will pounce upon copulating pairs, with the result that large balls of golden flies can be seen tumbling about the dung surface while the object of their desire is pushed and pulled in all directions; sometimes females are drowned in the dung surface or otherwise injured to the extent that they can no longer fly. When the density of males on and around pats is high, a male capturing an incoming female will carry her in flight to the surrounding grass to copulate before returning her to the dung to lay her eggs. During oviposition [egg-laying] the male remains mounted upon the female and pairs separate only after a clutch of eggs is laid."

Aside from making it clear that my friend is a man who truly loves his subjects of study, this lyrical description points out several crucial aspects of dung fly romance, and hints at why thinking outside the fertilization box will be illuminating. First, why would the males bother to take the females away from the melee before mating with them? Second, why bother staying while the female lays her eggs, which occurs after the male has deposited his sperm? And finally, why should mating take over half an hour, a seemingly excessively long time for the simple act of sperm meeting egg?

The first person to try to answer these questions was Geoff Parker, who in addition to being an evolutionary theorist is something of a dung fly devotee himself. He and others established that the males' behavior helps their sperm to compete with the sperm of any other males with whom the female mates. The last male to mate with a female typically fertilizes most of her eggs, particularly if he can stay engaged with her for at least 30 minutes and displace the sperm of her previous mates. This means that time spent hanging around the female or sequestering her from other males is time well spent, even if the male isn't actively engaged in transferring sperm.

After Parker's pioneering work, biologists threw themselves into an examination of the fate of sperm after mating, and hence into a scrutiny of the male organs themselves. There is nothing like a view of the genitalia of insects to convince you that the male equipment in human beings is rather dull and pedestrian in its appearance. In contrast, male damselflies have penis equivalents that boast a terrifying array of spikes, scoops, and hooks. The humble chicken flea has genitals bristling with strange knobs, kinks, and coils that Eberhard calls "one of the marvels of organic engineering," citing its "morphological exuberance." We never see these organs because the insects themselves are so small and their private parts are often held inside the body until they are needed, but similar well-cloaked monstrosities lurk in most insects.

What these elaborate structures do, more similar to the function of antlers on elk or the curving horns of bighorn sheep than to the genitals of many other animals, is fight with other males. The battles, however, take place while one opponent is completely absent, and the scoops and spines serve to remove a prior mate's sperm from the female's reproductive tract so that it can be replaced with the current male's ejaculate. Exactly what kinds of tools are needed depend on whether the rival's sperm is to be scooped out, poisoned, or merely drowned by a larger number of sperm. In some species, a male tamps down the sperm from previous matings, rendering it less accessible, before overlaying it with his own.

Sperm competition can also occur via the sperm itself and the chemicals that accompany it in the semen. Although they occur in many, perhaps all, insects, these chemicals have been best studied in the fruit fly *Drosophila,* which produces substances accompanying sperm that can kill the sperm of previous mates. These accessory proteins, as they are called, also influence the female's sexual behavior, sometimes rendering her less receptive to future matings, sometimes decreasing her overall life span but increasing the number of eggs she lays that are sired by her latest mate. Eberhard and his colleague Carlos Cordero call these seminal products *chemical genitalia,* because they can be seen as extensions of the more conventional physical reproductive organs. We are only just beginning to understand their complexity; Tracey Chapman from the University of East Anglia in the United Kingdom, in an article titled "The Soup in My Fly," referred to the bewildering diversity of seminal proteins as "an embarrassment of riches," surely the first time this phrase has been used in the context of sperm. At least 133 different substances have been identified, with doubtless more to follow. Whether each has a different function remains to be seen.

All else being equal, the more sperm that are present in an ejaculate, the more likely the male is to win at sperm competition, simply by overwhelming the prior male's efforts. In the Pieridae, a family of butterflies that includes the familiar cabbage white butterfly, ejaculates are significantly larger than in a family in which females are less likely to mate with multiple males, the ironically named (at least in this context) Satyridae. In insects, as in humans, sperm are produced in the testes, though insects generally lack external testicles housing the male organs. The larger the testes, the more sperm a male insect can produce, and you would therefore expect that sperm competition would cause the testes of species more likely to mate with multiple partners to evolve to a larger size than those in comparable but more monogamous species. That's exactly what was done by dissecting and weighing the testes in different types of Satyrids, and as expected, the greater the likelihood of females mating with many different males, the larger the testes relative to the size of the body.

Recently, sperm competition was actually experimentally shown to influence the evolution of testes size, not just indirectly via comparisons of species, in some elegant work by my friend Leigh Simmons and Paco Garcia-González, a Spanish scientist working in Leigh's laboratory at the University of Western Australia. Leigh has continued in the manure-inspired vein begun with the dung flies by performing pioneering research on dung beetles, those indefatigable insects that tidy up the world's ecosystems by removing the droppings of large mammals and using them to provide an all-purpose larder and nursery for the offspring. Many types of dung beetles occur all over the world, and in some species the males possess large horns used in combat with other members of their sex to gain access to underground tunnels excavated by females. Larger horns make it easier to win fights, but as with many insects, the sexual competition is not over after the physical battle is won. Some males sneak into the burrows of the winners and mate with the females behind the resident's back, as it were, and the only recourse of the former winner is to mate more frequently with the female.

One of the many obliging characteristics of dung beetles from the perspective of the scientists who study them is the ease of obtaining the raw material, so to speak. Leigh and Paco simply turn up at a local dairy farm and ask the farmer's permission to rummage around in the droppings left in the pasture, permission that is virtually always granted, albeit not without some quizzical looks. It is always difficult under such circumstances to decide exactly how much one should explain about the reason behind the request, striking a delicate balance between Too Much Information ("Here, let me tell you all about the evolution of male genitalia in beetles!") and sinister-seeming reticence ("Oh, nothing special, really. I'm doing a project on, um, sex."). From long experience, Leigh has figured out how to make such requests without alarming the farmers, and he and Paco duly brought back about a thousand beetles to his laboratory.

Leigh and Paco then performed what is called experimental evolution, by altering the environment of the beetles to see if the hypothesized selection pressure, namely, the risk of sperm competition, had the predicted effect on the beetles' testes size. It is really just artificial selection, the same process used to obtain do-

sider the female's side of the equation. Once Eberhard and a few others got things started, however, it became clear that female insects did not simply lie around passively waiting for the sperm to duke it out inside their bodies. One of the best examples of cryptic female choice is found in the humble flour beetle, the same tiny pest that infests the canisters of flour and other grain products in your kitchen. Lurking inside these miniature creatures is a hotbed of reproductive intrigue.

Tatyana Fedina, now at the University of Michigan, performed some extremely clever—if grisly—experiments on the beetles when she was a graduate student at Tufts University to see just how much say the females had over the fate of sperm. Flour beetle males pass sperm to females in a tube that turns inside out once it is in the female's body. The beetles mate with multiple partners, and some males father more offspring than others. But who controls paternity? Is it the males, via sperm competition, or the females, via selective use of sperm?

Fedina took advantage of the rather oblivious nature of male flour beetles when it comes to sexual activity and allowed them to either mate as per usual, or mate with a freshly killed female, something they did quite readily. She also starved some of the males so that they would seem to the females to be of poorer genetic quality than the others, and then compared how many sperm were transferred to the female's reproductive tract. Not surprisingly, the food-deprived beetles were less successful at transferring sperm—but only as long as they were mating with a live female. Males mating with the dead, and hence incapacitated, females showed no difference in their ability to inseminate, regardless of their condition. This means that the female herself must be doing something to influence the father of her offspring, favoring the well-fed and presumably higher-quality males.

Other, perhaps not quite so ruthless, experiments used anesthetized female flour beetles to study the degree to which females can control the movement of sperm inside their bodies. The immobilization of a female's musculature caused changes in the number of sperm inside different parts of her reproductive tract, further supporting the idea that females are more than simply vessels for sperm. A similar experiment was performed using a small moth; when a female moth was mated to two males, the larger individual always fathered most of the offspring, regardless of the order in which the matings took place. Again, this bias seems to be due to the female's actions, because anesthetized females showed no sperm in their sperm storage organs, even though the sperm themselves were just as mobile as ever, suggesting that the female has to actively shuttle sperm into the right place.

Among most vertebrates, sperm are deposited all at once during ejaculation, which means that the length of copulation probably has little effect on the number of sperm transferred. Many insects, however, transfer sperm in tiny packets, called spermatophores, that often attach to the outside of the female's body, leaving them perilously vulnerable to removal while the sperm are draining into the female's reproductive tract. Other species transfer sperm during the entire mating process, which can take many minutes or even hours. This means that if females control how long coupling lasts, they also control how many offspring a given male is likely to father. For example, female black field crickets in Australia let spermatophores remain attached longer for more attractive males (those singing more energetic songs) than for relatively wimpy males.

To induce females to allow spermatophores to remain attached, males in many different insect groups offer an enticement in the form of food. Several different kinds of male katydids produce not only the spermatophore like those of the crickets discussed above, but a nutritive blob attached to it called a spermatophylax. In some katydids, this structure takes several days to manufacture, and weighs a third or more of the male's body weight, representing a substantial offering. The female eats the spermatophylax, and its protein-rich contents enable her to lay more and larger eggs. The sperm are transferred to the female while she is eating the spermatophylax; when she has finished her meal, she often reaches around, breaks off the sperm-containing structure, and eats that too. The larger the spermatophylax, the longer it takes her to finish it, and therefore the more sperm enter her body.

Because the spermatophylax is so expensive to produce, each one represents a significant chunk of the male's mating effort for his lifetime. As a result, males in some katydid species become rather choosy about just who is entitled to receive one of the delectable morsels. Larger female katydids lay more eggs, which means more offspring sired by a male's sperm. Thus, as might be expected, in Mormon crickets (which are really katydids, not crickets, and which lack any religious affiliation so far as anyone can determine), males spurn small delicate females in favor of plump ones, a practice that may console failed dieters.

Other insects, such as hangingflies and scorpionflies, go out and catch prey items to present to females, who then consume the item while their hind ends are occupied with mating. Acquiring the prey items can be risky, since they are mainly obtained from spider webs, and so in a few species of scorpionfly, males offer specially produced wads of saliva to females instead. As with the katydids, the larger the gift, the longer the female will remain paired with the male. Sometimes, however, a male will simply grab a female and attempt to mate with her without offering one of these so-called nuptial gifts. Females take a dim view of such forceful behavior and generally won't stay coupled for very long with a male arriving sans offering. They also may be able to control the rate at which sperm are delivered into their reproductive tract. A recent study of a scorpionfly native to the Caucasus region in Europe found that while gift-bearing male scorpionflies remained coupled to females only twice as long as males attempting to force copulations, they transferred almost eleven times more sperm.

Females can also eject sperm after mating. Male damselflies and dragonflies, like the scorpionflies, simply grab females and mate with them, often removing the sperm of previous mates using the scoops and spines mentioned earlier. Most scientists studying these insects had assumed that the females had little control over who fathered their offspring, but a recent study by Alex Córdoba-Aguilar from the Institute of Ecology at the Autonomous University of Mexico showed that the females might have the last word. Córdoba-Aguilar noted that in many damselfly species, females had much less sperm in their storage organs than was present in the male's ejaculate. In fact, they seemed to have discarded so much sperm that they lacked sufficient numbers to fertilize all of their eggs. This seemed puzzling, or as he put it, "If females are using such sperm for oviposition, females are bad sperm administrators." He then collected females during different stages of the mating process and measured the volume of sperm present in their reproductive tracts. Then he counted the number of eggs that were laid after the females had mated with one or more males. It seemed that the females were favoring some males' sperm over others by ejecting the less-preferred males' contributions, long after the male himself had departed.

Males do seem to engage in some extreme antics to ensure that their ejaculates are not only placed in the appropriate part of the female's reproductive tract, but will be used by the female in fertilizing her eggs. As part of his ongoing studies of the intricacies of animal genitalia, Eberhard has uncovered some pretty racy stuff. A

2006 paper on spider mating behavior by Eberhard and colleagues Alfredo Peretti and R. Daniel Briceño, published in the ordinarily staid journal *Animal Behaviour,* contains passages that sound like what would happen if Danielle Steel were an entomologist: "Males squeezed females rhythmically with their enlarged, powerful genitalia throughout copulation." The title of the paper is no less suggestive, containing the words *copulatory dialogue,* again something one imagines those in the adult film industry to have mastered. In the spider under consideration, a rather modest-looking species called the short-bodied cellar spider, females "sing" during mating by moving their pedipalps, small appendages near the jaws, and making a sound the authors describe as "resembling squeaking leather." (If there is such a thing as spider porn, this is it.) As in many insects and their relatives, females mate with more than one male, and in this case the females seem to regulate paternity according to the ability of the male to be in tune with their wants and desires, if suggesting that spiders possess such things isn't too much of a stretch. The males adjust those rhythmic squeezes according to the sounds produced by the females, and males that were more responsive to females ended up fathering a larger proportion of the offspring.

My, What Big . . . Oh, Never Mind

HUMAN sperm cells have an easily recognized tadpole appearance and, while not exactly iconic in society, have their own modest place in kitschy, not to mention downright tasteless, objets d'art. There are neckties with stylized sperm cells, salt and pepper shakers, and of course the inevitable coin bank in the shape of a sperm cell (get it?). At the American Society for Reproductive Medicine's 2008 conference, sperm cell–shaped USB drives were handed out. But such aggrandizement aside, the truth is that human sperm cells have a pretty humdrum appearance compared with those of many insects.

In his original paper, Parker noted that sperm within an ejaculate must compete, not only with sperm from rival males, but with each other, and therefore any attribute making one individual cell better able to succeed should be subject to selection. This variability is particularly important in the insects, because sperm are usually not used to fertilize the egg immediately, as is the case in many other animals, but are stored for a period of weeks, months, or even years before they are used. This means that anything giving a sperm cell longer life or a competitive edge over the long term will be valuable. Indeed, insect sperm morphology is amazingly varied, including some with multiple flagellae, the whiplike organs used to propel the sperm through the medium. Sperm cells are much more variable across species than other kinds of cells, and even more variable than many body parts; one could, at least in theory, use sperm characteristics to distinguish species, the way that beak shape or feather color are used by bird-watchers to determine whether they are seeing a black-headed blue warbler or a white-eyed vireo. This is probably unlikely to catch on as a pastime ("Hey, guess what—I spotted a double-flagellated big head over the weekend!"), but it points to an often unconsidered source of biodiversity.

Some species of the humble fruit fly are the real sperm champions, at least if you think size is what matters. Male *Drosophila bifurca* look pretty much like any other fruit fly, namely, tiny and brown. But they have sperm cells that are about twenty times the length of the male producing them. To put this into perspective, for a human male six feet tall to achieve a similar feat, he would have to produce sperm cells that could span a sizeable portion of a football field, to carry on with the sports analogies that inevitably seem to accompany discussions of sperm competition. The cells are mostly tail and initially are in tangled coils resembling balls of yarn, so that the males employ what one scientist calls a "pea-shooter effect" to get the sperm transferred to the female.

Needless to say, manufacturing such behemoths is energetically costly, and a male can't produce nearly as many so-called giant sperm as the ordinary variety. Male *D. bifurca* therefore "use their sperm with female-like judiciousness," according to Scott Pitnick from the State University of New York at Syracuse. Unlike most species of insects, including other types of *Drosophila, D. bifurca* males and females mate with roughly similar numbers of partners. The precious cells are produced on demand, with more being manufactured when males are given greater access to females and fewer when mating opportunities are scarce.

The function of the elongated tails is unclear. Some researchers suggest that they may block other males' sperm from getting through the female's reproductive tract. Alternatively, large sperm may have evolved because of selection by females for the more exaggerated forms of the cells, making them what Pitnick and his colleague Gary Miller called "the cellular equivalent of the peacock's tail." Pitnick and Miller took laboratory populations of *D. melanogaster,* a more commonly used fruit fly than the giant sperm–bearing *D. bifurca,* and subjected them to artificial selection experiments similar to the ones that Simmons employed with the dung beetles. Here, instead of constraining the number of mates an individual had, Pitnick and Miller selected directly for either increased or decreased sperm length or the length of the females' primary sperm storage organ, the seminal receptacle. Like most insects, female fruit flies have convoluted organs used to keep the sperm until the eggs are fertilized, often many days later.

After thirty or more generations of the selection treatment, flies from the different groups were mated to each other, and the relative success of the different types of males at fathering offspring was calculated. The newly created long-sperm males were much better than the short or normal length sperm males at fertilizing eggs of the females with longer seminal receptacles. When females had short seminal receptacles, sperm length didn't matter. Pitnick and Miller concluded that the giant-sized sperm evolved because the female reproductive tract selectively biases paternity in favor of males with longer sperm. What caused the female's seminal receptacle to become longer — and why *D. bifurca* is the fly equivalent of a bird of paradise, while other species are the drab sperm sparrows of the *Drosophila* world — isn't clear.

Complicating the story is the finding that among at least some other insects, such as the dung beetles, shorter sperm seem to do better than longer sperm. Male dung beetles in better condition, with better nutrition as larvae, produce shorter sperm. And fathers that sired sons producing short sperm also had daughters with larger sperm storage organs. At least with regard to sperm, size may matter, but it isn't always better to be big.

A Caste of Thousands

AMONG virtually all butterfly species and some other insect groups, two types of sperm, sometimes called castes, as in the worker and queen castes of social insects, are produced: eusperm, which has a DNA-carrying nucleus and is capable of fertilization, and parasperm, which is smaller and has no genetic material. Some scientists have suggested that the different sperm morphs have different functions, with only a tiny minority of sperm actually able to reach the egg. The other sperm cells act as blockers of rivals or helpers of the real champs (for example, they make it easier for the fertilizing sperm to move through the female reproductive tract) but are themselves sacrificing their own chances for survival. Some years ago it was suggested that similar divisions of labor occurred in human sperm cells, and the nonfertilizing sperm were dubbed kamikaze sperm, for obvious reasons.

It is a colorful theory, but the evidence, at least as it pertains to humans, is weak at best. In mammals, many sperm that appear nonfunctional didn't get that way through a plan; they represent errors in manufacturing. And the evidence about what is retained versus rejected by women's reproductive tracts comes from a very few studies of what is exuded after sex, using samples provided by a group of volunteers who may or may not represent the general population.

Insects, though, are another story as far as sacrificial sperm goes. In butterflies, the theory that seems to have the most support is that the nonnucleated sperm cells are cheaper to produce and, hence, may act as "filler," allowing the male to swamp out other ejaculates with quantity if not quality. The more likely a butterfly species is to experience sperm competition, the longer the eusperm. But the same function does not seem to occur in the other insects with two kinds of sperm, and until recently the role of these odd self-sacrificing cells was a mystery.

Some recent work by Luke Holman and Rhonda Snook at the University of Sheffield in the United Kingdom suggests that looking at the situation from a female perspective may help explain the evolution of this obviously masculine trait. They used yet another species of *Drosophila, D. pseudoobscura,* which also has the two sperm types, to see whether female flies might be the ones controlling the situation. Indeed, many of the sperm are actually killed by the chemicals or cells present in the female's reproductive tract, and the DNA-containing eusperm seem to be particularly sensitive. When males produced more of the parasperm, the eusperm were protected from the spermicidal activity. The parasperm appeared to be acting as shields for their more fertile brothers. A few authors actually refer to the parasperm as *soldier sperm,* but I think that presumes that they are fighting each other, when as Holman and Snook point out, the female seems to be playing an active role in their demise.

Why should the female reproductive tract be such an unwelcoming environment for sperm? Holman and Snook speculate that one possibility is that females are using their own reproductive organs as a screening device, making the sperm from various males run (or swim) a gauntlet of tests before being allowed to fertilize the eggs. In other words, cryptic female choice could explain the evolution of a seemingly nonfunctional cell type. The criteria that the females might use to distinguish among suitors are not well understood. It may be that, as with the exceptionally long tails of the giant sperm in *D. bifurca,* parasperm are a kind of secondary sexual trait, like the peacock tail.

An answer to the question of why sperm are so variable, as well as how likely it is for females to sort through sperm in their reproductive tracts and the function of all those chemicals in the semen, will likely require examining the situation from the sides of both males and females. The fierce activity that occurs inside a female after copulation does, however, provide a possible insight into that sex difference in postcoital behavior. With all that commotion going on in there, who could sleep?

CHAPTER 6

So Two Fruit Flies Go into a Bar . . .

As SOMEONE who works on sexual behavior in animals, I've grown used to getting a lot of off-the-wall questions from curious members of the general public. Topping the list is homosexuality and whether it occurs in species other than our own. (Another inexplicably popular area of inquiry is whether animals exhibit oral sex. I still have no idea why people want to know the answer to this, and have always been afraid to ask.) And any media mention of homosexual behavior in animals always garners lurid headlines and stimulates acrimonious online debates. In 2007, for instance, news that scientists induced homosexual courtship in male fruit flies by changing the levels of a chemical that is key to many processes in the nervous system was greeted with predictable tabloid hyperbole: "Scientists make fruit flies gay, then straight again." On science and gay rights blogs alike, discussion raged about whether this meant that a drug altering sexual orientation would, or should, be developed by the demon Big Pharma. Others trotted out well-worn arguments about whether sexual orientation is learned or genetic and about its existence elsewhere in the animal kingdom, and then meandered into why places with large contingents of gays — such as San Francisco and Boulder, Colorado — are often such nice places to live.

Similarly, every time the issue of gay marriage rears its head, animal homosexuality comes up, in part because arguments against gay marriage often invoke phrases such as "natural order," "natural law," or "crime against nature," which make it, well, natural to wonder about whether birds — and even the bees — do that, too. And marriage aside, animals have always featured more generally in discussions of how "natural" homosexuality in humans might be, although which side their behavior is used to support differs. On the one hand, some gay activists have pointed to the widespread occurrence of same-sex courtship among animals from penguins to whales as evidence of it being part of the natural spectrum of acceptable behaviors. Animals are also sometimes used to bolster the argument that sexual orientation is not a choice but a genetically influenced, or even genetically determined, trait. Some conservatives, on the other hand, feel that animals exhibiting a distasteful behavior just underlines its debased nature. Simon LeVay, a prominent researcher in the genetics of sexual orientation, throws up his hands: "The question of whether animals engage in same-sex sexual behavior has been debated for centuries, most often in the context of efforts to stigmatize homosexuality. Three classes of answers have generally been offered: 'Animals don't do it, therefore it's unnatural'; 'Animals *do* do it, therefore it's bestial'; and 'Some animals do it, and those are the unclean animals.'"

At some point in the argument someone inevitably says, as if no one else would have ever thought of it, that animals do all kinds of things we don't want to emulate, for example, eating their young or abandoning their elders. The implication, presumably, is that what animals do is sometimes repugnant, so we should ignore their behavior when considering our own. While it's certainly true that we don't need to use other species as role models for behavior to emulate, animals need not mirror all aspects of our lives to be useful in teaching us about some of them. We use animals as experimental models for many parts of our biology that they do not possess in their entirety. We can learn a great deal about how babies grow into adults by observing rats, even though rats never learn to drive or go to college. And we are fascinated with the things that animals do that seem so uncannily similar to what we do, as anyone who has watched a mother monkey expertly sling her baby on her hip before setting off for a nearby shrub can testify.

Insects play a special role in our use of animals to help us understand ourselves, as I argue throughout this book. Because they are rarely cared for by their parents, and usually live relatively solitary lives without the input of others, the behavior they exhibit as adults is largely controlled by their genes. And although we are increasingly discovering how flexible their behavior can be, as I discuss in the chapters on insect learning and personality, it's still a safe bet that if a bug seems to be homosexual, it didn't get that way because of an absent father or overbearing mother messing things up during its larval stage. Their behavior is thus stripped down to its essentials, a handy tool for looking at complex actions.

So what do we know about homosexuality in animals, particularly insects? And what does that tell us about sexual orientation in humans? The results of studies showing same-sex behavior in flies, beetles, and butterflies are coming in every day. This news is significant for several reasons, but it is meaningless for another one, and that one is the reason that many people are interested in it in the first place.

Lowering of Moral Standards in Butterflies

As Bruce Bagemihl points out in his 1999 book *Biological Exuberance,* researchers have been noticing same-sex behavior in both wild and captive animals for many years. This is not to say they were always happy about it or viewed their discoveries dispassionately; a scientist greeted the sight of male bighorn sheep mounting each other and forming long-term homosexual bonds with: "I still cringe at the memory of seeing old D-ram mount S-ram repeatedly. . . . To state that the males had evolved a homosexual society was emotionally beyond me. To conceive of these magnificent beasts as 'queers'—Oh, God!"

Even lowly invertebrates are subjected to such dramatic responses; the 1987 volume of the *Entomologist's Record and Journal of Variation* contains "A Note on the Apparent Lowering of Moral Standards in the Lepidoptera [butterflies and moths]," a gem I have read several times, still without being sure whether it is meant to be tongue in cheek. In it, the author laments, "It is a sad sign of our times that the National newspapers are all too often packed with the lurid details of declining moral standards and of horrific sexual offences committed by our fellow *Homo sapiens;* perhaps it is also a sign of the times that the entomological literature appears of late to be heading in a similar direction." He then goes on to detail observations of male Mazarine blue butterflies, a lovely European species, vigorously and persistently courting other males, particularly when the object of their attentions had just emerged from the chrysalis, despite the ready availability of females. The note concludes with the reassurance that several heterosexual pairs—referred to as "normal"—were also seen, thus indicating that "at least some individuals had the furtherance of the colony at heart and the appearance of the colony next year is thereby assured."

Admittedly, *The Entomologist's Record* is not the most prestigious or widely cited of journals and contains quite a few other anthropomorphic articles, such as the poignantly titled, "Do Copper Underwings (*Amphipyra* spp.) Crawl Away in Order to Die in Peace?" Nevertheless, same-sex behavior in animals, whether sheep or butterflies, seems to bring out this kind of histrionic reaction in those who observe it. And Bagemihl points out that we are probably seeing only the tip of the homosexual iceberg, since many more researchers may be seeing similar behavior in their study organisms but ignoring it or dismissing it as a meaningless aberration.

Because insects do not invite the same identification or anthropomorphism as mammals and birds, though, we can at least hope to use them as testing grounds for our ideas without automatically falling back on our biases. Most modern scientists would dismiss the idea that moral standards exist at all in butterflies, much less that same-sex behavior is a sign of them. What kinds of homosexual behavior do we see in insects and other invertebrates?

For example, the males and females of a small spider that biologist Rosemary Gillespie studied in Hawaii do not exhibit any elaborate courtship behavior before mating. Instead, they simply leap at each other, fangs outstretched. If such abrupt amorousness is acceptable to both parties, the fangs become locked together (giving new meaning to the phrase "hooking up"), and the female curls her abdomen around so that the male can insert his sperm-bearing organ into her reproductive opening. A captive pair of males that Gillespie had collected a few weeks earlier exhibited much the same behavior in their container, remaining coupled for 17 minutes. Similar same-sex pairings, usually between males, have been seen in captive and wild beetles, locusts, wasps, and a kind of fly that lives near streams and lays eggs in water lilies.

In the blue-tailed damselfly, females come in three colors, one of which resembles that of males. Hans Van Gossum and his colleagues at the University of Antwerp in Belgium kept male damselflies either with other males or in mixed-sex groups and then allowed the males to choose between a female or another male in a small cage. Males that had experienced the damselfly equivalent of a British boarding school were more likely to then seek out another male and form a pair with him, while the males from the coed environment were more likely to pair up with a female.

To interpret this puzzling result, we need to know some details about the sex lives of this insect group. Mating in damselflies and dragonflies is both distinctive and complex, and because the insects are aerial, one can often see mated pairs flying over the surface of a stream. A male damselfly or dragonfly, unlike other insects (and most other animals, for that matter), actually has two sets of genitalia, one at the tip of his abdomen and the other closer to the center of his body, at the underside of the second abdomen segment. Before mating, the male transfers his sperm from the tip to the more central location. Then, once a male sees a female he intends to mate with, he flies up and grasps her behind the head with his rear appendages in what is called a tandem position. The pair may fly together like this for several minutes or even longer. Eventually, if the female does not reject the male, they land on a plant or some other object and form a wheel: the female bends her abdomen tip to reach the male's secondary genitalia so that he can transport his sperm into her reproductive tract. (I've always thought the wheel looks a lot like a heart, and often show images of paired damselflies to my class on Valentine's Day.) The couple will remain in the wheel position for up to 15 minutes, and the male often accompanies the female after mating is over, still in the tandem position, while the female returns to the water to lay her eggs.

This rather convoluted process means that males can potentially waste a considerable amount of the females' time by persistently grabbing them in midair and chasing them around a pond or stream. Given that the females live only a few days or weeks in most species, and that it's important to find the best place and time to lay eggs, such harassment is more than merely annoying — it can compromise the female's ability to reproduce. The malelike morph of the females in the blue-tailed damselfly and several other species is thought to have evolved to allow the masquerading females to avoid some of the pestering, because males initially at least mistake them for other males and are less likely to bother them. This in turn means that selection will act on the males to make their acceptance rules for whom to court and who to shun a bit more flexible, so as not to miss out on any mating opportunities.

Van Gossum suspects that having this relatively open-minded response to all members of the same species, regardless of sex, means that a certain proportion of male-male pairs is inevitable, even though they obviously cannot increase the reproductive success of the males involved. The idea is that it's better to have a coarse decision threshold and risk accepting some mistakes than to be more discerning and risk missing some actual females. It's a bit like testing for certain cancers, where doctors would rather put some people through unnecessary biopsies and anxiety for a false positive than risk missing some actual signs of disease. In evolution, as in medicine, where to set the bar is not always clear.

Boys Will Be Girls Will Be Boys, Naturally

PEOPLE sometimes conclude from this kind of work that the damselflies, or Gillespie's spiders, or any of the other insects and their kin observed in homosexual pairings are making a mistake, and therefore human homosexuals are likewise in error, some kind of evolutionary fluke. Interestingly, a *National Geographic* story on Van Gossum's study suggested, "Such flexibility may also lead to genuinely homosexual damselflies." This implies, I suppose, that the damselflies in the Belgian experiments were somehow not really gay, although it's hard to know what the litmus test might be.

Instead, I think it makes more sense to see the flexibility in mating behavior, same-sex courtship and all, as part of the animals' natural repertoire. We cluck disapprovingly over the males' supposed errors, but that represents our misunderstanding of how evolution works. François Jacob famously said that nature is a tinkerer and not an engineer. What he meant was that natural selection doesn't produce perfection; it produces traits that are good enough. We often think of this in connection with our bodies, so that we have spines that are not really adapted to walking upright or immune systems that sometimes overreact to give us allergies to harmless substances, but the slop is part of every system, including behavior.

The chrysanthemum longicorn beetle (*Phytoecia rufiventris*) is a lovely insect with a ruby red spot on its back and a rust-colored abdomen. It is a pest of chrysanthemums, as the name suggests, and a single female can kill as many as seventy plants by laying her eggs in the stems, which makes understanding the beetles' biology of interest to horticulturalists. Unlike many insects, the chrysanthemum beetle lacks sex pheromones, those come-hither odors often employed as long-distance mate attractants and sex identifiers. The sexes find each other in the first place because both males and females are attracted to plants of a certain height. Qiao Wang at Massey University in New Zealand and his coworkers discovered that the male beetles reacted similarly to males and females when they first encountered them on a plant stem by attempting to copulate. After the male mounts, he engages in a rather complicated and lengthy probing with his abdomen until he can touch a tiny segment of abdomen of the beetle underneath him, and it is only at that point that he can determine whether he has mounted a male or a female. Eventually he disengages from an individual found to be another male, but Wang and his colleagues suggest that "males may 'waste' a lot of time during their reproductive life."

But time wasted is in the eye, or maybe the pheromone glands, of the beholder. Sure, if the beetles had a more foolproof way to determine who was who, they would have more time to feed, or hide from predators, or do crossword puzzles for that matter (what's a nine-letter word for "life-destroying chemical"?). Similarly, if human beings had pelvic girdles that could more easily accommodate a full-term fetus, childbirth would be a breeze and the militant advocates of natural versus medically assisted labor would have to find something else to clash about. But in both cases, evolution didn't produce the best solution, it produced what worked.

Seemingly maladaptive traits may persist because no genes for a more efficient matchmaking technique or a less painful birth process exist for natural selection to act upon. If a male beetle with a genetic mutation allowing him to sniff out females were to crop up, he might be wildly successful, and in time his progeny would outnumber the old models. Maybe someday that will happen, thanks to the vagaries of genetics. In the meantime, as long as we have chrysanthemums for them to plunder, the beetles muddle along. Alternatively, the trait might represent a compromise between competing selection pressures: you can have a good pheromone system, but then your predators can find you, or you can't perform some other essential task. A more capacious pelvis might come at the expense of walking. If the fetus were smaller when it was born, a solution many of our primate relatives opt for, we'd have babies with less brain power. Ironically enough, entomologists often exploit the pheromones of insects by constructing traps that emit an artificial version of that enticing odor that ordinarily means that romance beckons; when the hopeful suitors arrive, they are summarily dispatched. The lack of sex pheromones in the chrysanthemum beetles makes them that much more difficult to lure to their deaths.

A similar delay in figuring out who is male and who is female occurs in a species with one of the most grisly mating habits in the world, the African bat bug. These insects are related to bedbugs, although as the name implies, they ordinarily suck blood from bats in their caves rather than human beings in their beds. Both insects reproduce not via males depositing sperm in the female's reproductive tract but through a process called, accurately enough, *traumatic insemination.* The male literally pierces the body wall of the female and leaves his sperm to swim through her body cavity to fertilize her eggs. Males always stab the female's body in the same place, and when they do so their organ must pass through a specialized structure in the females that is unique to this insect group. This structure helps to protect them from the onslaught of bacteria and other nasty material that is introduced into the female's body with the sperm. The bugs seem to be unable to distinguish males from females until rather far along in this process, which means that a certain proportion of the time, males will attempt to mate with other males. Bat bug males, it turns out, also have structures at the wound site that are similar to those possessed by females, although these differ in some anatomical details. Scientists from the University of Sheffield in the United Kingdom who study the insects speculate that the males evolved these structures to signal other males that they are not females, and perhaps even to provide some protection from pathogens in the event that the male doesn't get the message. Here too, selection on males to be profligate with their mating attempts may simply have overruled selection to be more reserved, even with some unwanted side effects.

So a certain amount of homosexual behavior isn't any more of a mistake in the beetles than any other trade-off between two traits might be. Instead, that very flexibility in mating behavior, where decisions are made based on shifting criteria that may be apparent only at the last moment, might itself be favored by natural selection. It allows animals to be opportunistic in their behavior, and increases their ability to roll with the punches of a changing environment.

Mistaken identity, however, doesn't seem to be what's going on in a tiny fly that lives on the water lilies of English streams. Males wander on the surface of the leaves, pouncing on anything that remotely resembles a female and a few things that do not, such as gray specks of decay on the plants, or flies of other species. After he succeeds in mounting a female, the pair embarks on an elaborate courtship ritual in which they rock back and forth for up to 15 minutes. An uncooperative female quickly puts a stop to this activity, in which case the male leaves without bothering her further. Sometimes, however, a male mounts another male, and in these cases the mounted individual vigorously resists the overtures while the mounting male clings to his back as if to a tiny bucking bronco. Ken Preston-Mafham, who has studied the flies in Warwickshire, believes that the mounting male is preventing his partner from getting to the females that will light upon the lily leaf. If males are competing for access to the females, a male that simply rides another individual is in the best position to leap off his rival and seize the female himself.

Finally, there may be some unforeseen advantages to homosexual behavior, regardless of why and how it arose. As I mentioned, flour beetles, the tiny pests infesting your kitchen cabinets, are useful models for genetic and other biological research. Like the other insects I just described, male flour beetles will mate with other males. Work in Sara Lewis's laboratory at Tufts University in Massachusetts showed that when one of the males mated with a female right after such a homosexual interaction, on a few occasions enough sperm from the other male was left that it actually fertilized some of the female's eggs. Although this is unlikely to be a frequent occurrence, it suggests that some reproductive benefit could partially offset any wasted time or effort in the male-male interactions.

Flies with Designer Gay Genes

WEIRD sexual proclivities of bedbugs aside, what people really want to know is whether homosexuality has a genetic basis. Because it is much easier to search for such genes using quick-breeding study animals, the fruit fly *Drosophila* has become the poster insect for studies of sexual orientation genetics, as it has for so many other traits. Although people rarely identify with insects, particularly tiny buzzy ones such as flies, in this case the media has reliably been all over any new finding that deals with homosexuality in *Drosophila,* with headlines such as "Fruitflies Tap in to Their Gay Side," "Gay Drunk Fruit Flies," and even "Gay Fruit Fly for President" (not sure what that was about, frankly). Google "gay fruit flies" and you get upward of 270,000 hits.

Scientists were not looking for homosexual flies when they began this research. Indeed, most if not all of the researchers whose papers end up providing fodder for headlines like those above would not describe themselves as studying sexual orientation at all.

Instead, they are trying to understand how the brain sends and receives signals from the sense organs, or attempting to break down the processes of courtship and mating into their most fundamental components. What exactly has to happen for boy to meet girl so that baby can make three (or thirty, or maybe three hundred, in the case of the flies)?

It turns out that sex, even for such relatively simple animals, requires a sophisticated orchestration of steps. Although different species of *Drosophila* do things somewhat differently, in many fruit flies the females must go to a specific kind of rotting fruit or other plant matter to lay their eggs. When they are there, the males detect the presence of a potential mate by smelling the surface chemicals on the female's body, and then pursue the object of their affection, performing a stereotyped series of movements, attempting to lick the female, and vibrating their wings to produce a song that is audible to human ears only if it is greatly amplified. The details of the movements and the song vary among species, and males differ in how vigorously they perform the actions and in the way in which their advances are received. Often the female walks away or lashes out with her legs in response. If a male is successful, the female stops long enough to allow him to mount her and transfer his sperm. Both in the laboratory and in the wild, males will also attempt to court other males, particularly younger ones that have just become adult.

One can, of course, study the flies simply by observing their behavior using a magnifying glass to get at some of the finer details, but for many years scientists have been using extremely sophisticated genetic technology to understand exactly which genes control which aspect of the mating ritual, and how they interact. It is now possible to produce knockout strains of the flies, which lack a particular gene but are otherwise like normal, or wild-type, as they are called, *Drosophila*. Alternatively, scientists can manipulate individual genes so that they are still present but are inactivated; genes can also be inserted into places they wouldn't normally occur.

One of the most important genes regulating sexual behavior in the flies is called *fruitless* (many genes in model organisms have special names, some of which are quite fanciful, for example, *sonic hedgehog*). Flies with one kind of mutation in this gene will try to court females, but they do so incorrectly. It's still not clear where their problem lies, but it may be that they fail to fully extend their wings to sing, a deal breaker from the female's perspective. This defect applies only to courtship—the mutants can fly normally and can flick their wings dismissively when rejecting advances made by another male. Flies with other types of mutations of the *fruitless* gene court both males and females. When several of the mutant males are placed in a Petri dish, they form male-male courtship chains in which each male is simultaneously both courting and being courted. Female flies with the altered *fruitless* gene will court other females with the same stereotyped set of movements ordinarily used by males.

The *fruitless* gene affects many different parts of the fly brain, each of which is important in regulating sexual behavior. A Japanese researcher, Ken-Ichi Kimura, meticulously dissected the brains of *Drosophila* that did and did not have the mutation in *fruitless*. He and his coworkers found that a just a handful of nerve cells in the wild-type males are absent in the males with the mutation. In the females that court other females, the cluster is also present, although normal females lack it.

So is *fruitless* "the gay gene," or do the nerve cells themselves keep flies from being gay? Not so fast. Kimura and his colleagues also worked with mutations on another gene, called *doublesex*.

They found that a nerve cell group that is affected by mutations on both of the genes simultaneously can turn on courtship behavior in females. Ordinarily, this cell cluster dies in females because of a feminizing protein in the brain, but if *fruitless* is present, the cluster survives. Both of the genes are needed to ensure that males court females and females don't.

Then are both genes "gay genes"? Once again, no. Just having the genes that control the courtship behavior itself isn't enough. A male also needs to distinguish that a female is out there in the first place, which means processing the sight, smell, and maybe sound of another fly, and yet more genes seem to be involved in that process. The male flies' behavior is triggered by pheromones, or odors that are emitted by the female. Females that have already mated, and hence are more likely to reject the courtship advances of subsequent males, produce a different odor than virgin females or other males. But the male also needs a gene to enable him to detect those chemicals. Flies don't have separate taste and smell organs; they detect both with sensory cells on their feet (which is why they often walk on things before deciding whether they are food, and why they can spread germs so easily), and work by researchers at Duke University published in 2008 showed that a pheromone detection gene was critical to the mating game. This one is unimaginatively dubbed *Gr68a,* and males with a mutation in it will court already-mated females as well as other males, and they will even go beyond the tapping and singing behavior to try and actually mate with the males. What is more, the signals that it receives bypass the rest of the nervous system and go directly to the brain. Such an express route is unusual for a smell or taste receptor, which underlines the essential nature of *Gr68a* to the mating process.

Other flies that exhibit male-male courtship have alterations in genes called *dissatisfaction, prospero,* and *quick-to-court.* What's more, the neurochemical dopamine, which is important in a wide variety of physiological activities, including learning, movement, and the brain's processing of painful or pleasurable stimuli, also turns out to feature in same-sex courtship in *Drosophila.* Dopamine is found in many animals, both vertebrate and invertebrate, including humans, but if you increase it in the flies, males are more likely to court other males, although they don't change how they react to virgin females or to odor cues in general. And if news about dopamine alone leaves you cold, further research in this area demonstrated that when flies genetically altered to be unable to release dopamine at normal temperature were exposed to ethanol, the type of alcohol in beer, vodka, and other adult beverages, they too exhibited same-sex courtship. The male-male courtship became more pronounced with repeated exposure to the alcohol; the experimental arena where the scientists placed the flies was quickly named the "Flypub," and the inevitable news coverage trumpeted, "Fruit Flies Prove That Alcohol Makes People Gay."

Better Sex through Chemistry

MOST of this research used flies that exhibited the altered behavior permanently. But one of the most exciting new developments in the genetics of *Drosophila* sexual behavior showed that the tendency to court males or females could be switched on or off within minutes.

Dave Featherstone at the University of Illinois in Chicago said in an email to me that he was envious of what I do, because he "got into biology because I imagined myself traveling all over the world living in the wilds watching animals. Somehow I ended up studying bizarre minutia in a lab. I might as well be an accountant." As someone who does watch animals in nature for a living, at least some of the time, I was flattered by his comment, but his modesty underplays the significance of his work, which is hardly bizarre or trivial. Featherstone is interested in how the cells of the nervous system send and receive messages, particularly across the gaps between them, called synapses. His laboratory focuses on a nervous system chemical called glutamate, which his website describes as follows: "Glutamate is the voice by which brain cells speak to each other. Glutamate receptors are the ears by which they hear."

Information—whether about sex, food, or anything else—does not simply slosh from one nerve cell to another, making its way haphazardly to the brain. Instead, the receptors regulate which memories are retained, which behaviors are executed, and which signals are recognized as important. So Featherstone studies glutamate and its role in brain messages using *Drosophila,* which show many of the same patterns of glutamate use as humans but are obviously much easier to manipulate.

For the work on the chemical courtship switch, Featherstone and his colleagues used male flies with a form of yet another gene called *genderblind.* Males with a mutation of that gene, like the *fruitless* mutants, will court other males as well as females. This is simple observation—what is exciting is that the scientists went on to pin down why. The *genderblind* gene controls the transport of glutamate out of glial cells, which are nervous system cells that do not conduct electrical signals themselves but communicate with and support other cells. Glutamate in turn can control the synapses, those junctions between other nerve cells, and synapse strength is important in determining many aspects of behavior. By altering synapse strength either genetically or chemically, independent of the mutation, the researchers could alter, sometimes within minutes, whether the flies would court males as well as females, or only females. Then they could be switched back again.

The altered males interpreted the pheromones of other flies differently than their wild counterparts because they had too many glutamate receptors at the junctions between the nerve cells. What would ordinarily be a male smell that induced other males to keep away was perceived by the mutants as stimulating.

What exactly was going on? Recall that even wild-type male *Drosophila* will court other males, particularly when they have recently become sexually mature. The courted males reject them, and they learn to stop trying, generally within about half an hour of getting nowhere. But the *genderblind* flies just don't take no for an answer, which suggested to Featherstone that what was really happening was at least partly a failure to learn from experience. Current research in his lab is focusing on how glutamate is involved in this learning process. Ironically, then, the search to find genes and chemicals associated with courtship and mating led to the discovery that learning, that most plastic of behaviors, is at the heart of the matter. Nothing could be farther from demonstrating that a gene or genes causes homosexuality.

Regardless, the media fell on Featherstone's discovery like gay-gene-seeking jackals. Perhaps because of the rapidity of the switch, commentators seemed to think that the work indicated it was possible, even likely, that scientists could develop a pill that would alter human sexual orientation, and hence be used to "cure" homosexuality, or that said drug could be used recreationally, so that one could be gay under some circumstances and straight under others. When I wrote a brief op-ed piece about Featherstone's work for the *Los Angeles Times,* in which I noted that the really interesting part of the research, the role of glutamate, had gotten ignored in all the brouhaha, I got emails accusing me of promoting genocide of homosexuals. Another article's title demanded, "If There Was a Gay-Straight Switch, Would You Switch?" Never mind that not a single answer to the question appeared in the article itself. Even some of Featherstone's colleagues questioned his use of the word *homosexual* in his paper on the work, calling it "tabloid language."

Aside from the fact that anyone who thinks that "homosexual" is tabloid language hasn't been spending nearly enough time at the grocery store checkout, much of this reaction was seriously off base. Featherstone points out that his work looked at courtship behavior that was indeed directed at members of the same sex, hence homosexual. But many people use the word to talk about sexual orientation, or the preference for one kind of partner over another. As he says, "Our data as well as recent data from mice suggest that mate choice is not some sort of 'compass arrow' that can only point at one target. . . . Let me make an analogy: mate choice is a lot like food choice. The fact that I like corn dogs doesn't keep me from liking pizza. They are separate sets of sensory stimuli, to which I can respond independently. . . . 'Homosexual' and 'heterosexual' are simply descriptive terms that define particular types of mate choice, same as 'corn dog' and 'pizza' define particular types of food."

This still doesn't suggest that in humans, sexual preferences are as easily manipulated, or that we become gay or straight with the same facility with which we choose our lunch menu. (Though, really, Dave—corn dogs?) But it does illustrate how easy it is to have this research misinterpreted by scientists and nonscientists alike.

Taking a Pill?

So what does the research from insects tell us about homosexuality? All of the scientists using genetic alterations in *Drosophila* hasten to point out that despite the flies sharing 75 percent of human disease genes, no counterpart to *fruitless* exists in people. So even though what the flies are doing looks at least somewhat like what humans do, the insects got to a similar destination through vastly different modes of transportation on different highways.

It is certainly true that the attraction to members of one's own sex is common in nature among many species, and its sources can be traced in the lab, at least for the flies. If that resonates with your world view on homosexuality, whether to accept or eschew it, so be it. But homosexual behavior means something different to the flies than it does to more complex and more social animals, such as the primates, birds, and other vertebrates that exhibit same-sex behavior. For example, Laysan albatross in Hawaii form female-female pairs that stay together for many breeding seasons, rearing chicks together if one or both of them has been inseminated by a male in the colony. Bonobos, smaller relatives of the chimpanzee, frequently exhibit sexual behavior between males or between females; sex seems to be used in bonobo society as a way to resolve social tension in the group. In these and many other animals, sexual behavior is about more than reproduction. People unfamiliar with life in the wild often envision animals keeping their sexual contact to a businesslike procreative minimum, where male and female meet, mate, and part as soon as the plumbing has everything lined up. But in social animals, sex is not just reproduction—it is communication, part of a continuum of dealing with other members of your species.

Fruit flies and the other insects I have been discussing do not have elaborate social systems in which such subtleties are important. Yet they still exhibit same-sex courtship and even mating. The conclusion, though, is not that bugs are stupid, but that sex is hard. Figuring out how to do it involves a complex interplay between genes and the environment. Featherstone and many of the other researchers using genetically modified flies measure their behavior by observing how the mutants respond, not to a living, breathing companion, but to a male or female with its head cut off. The fly's body still emits the same odor cues and provides the same audience for the displays, but it obviously cannot interact with its partner. The decapitation controls for the inevitable interaction between individuals that could alter the results. The genes don't just issue commands that make the flies behave in a certain zombielike way regardless of circumstances. Instead, the genes, and the chemicals they deploy, affect the way that experiences such as being rejected or accepted by another fly are interpreted.

The flies and other insects may also use some of the same-sex interactions as a way to practice their technique. Young male *Drosophila* are often courted by older ones, and Scott McRobert and Laurie Tompkins showed that males that had been the recipient of such courtship in the lab were more successful in wooing females later in life. The difference was not huge, but in evolution, every little bit counts. In a different species of *Drosophila* than the one used for most of the genetic research described above, males that are isolated from other flies during development have a hard time telling their own species from similar ones, a crucial skill, since hybrid offspring are not fertile and, hence, an evolutionary disaster. It takes experience with one to know one, it seems, even though *Drosophila* lack a true social organization like that of wolves, bonobos, or even bees. These flies also exhibit male-male courtship, with the socially ignorant males showing it much more frequently than males reared in the company of other flies.

Part of what seemed to get the public so excited about Featherstone's work was the idea that it wasn't a gene, but "a chemical" that altered the sexual behavior of the flies, which led to the speculation about popping a pill to change one's sexual orientation. If what's sauce for the fly is sauce for the human, this could mean that chemicals in our own nervous systems are involved with sexual orientation, too. But I don't find this in the least alarming, or indicative of some dystopian possibility of transforming people from gay to straight and back again. The truth is that chemicals no more control who we are sexually attracted to than they do anything else. Which is to say, everything and nothing.

Chemicals are where the body's rubber meets its road. They are how our genes exert their influence. It's fine to say that a gene controls eye color, or digestive speed, or whether we like mangoes, but what does that mean? Somewhere, a chemical is involved. Better living through chemistry? It's more like living through chemistry, period. Featherstone's lab has winkled the details out of the devil by connecting the gene to the proteins it codes for and the action of the substances those proteins control. None of that alters the crucial role of experience, even in a creature such as the fruit fly.

What does Featherstone himself want to do with this information? He doesn't seem motivated to get in the pocket of Big Pharma and develop a drug to enable people to go from straight to gay. Presumably he recognizes that this isn't possible. But he has some ideas. Going back to his website, "An understanding of *Drosophila* neuroscience raises the possibility that we may be able to engineer a ruthless bionic insect army, and use it to take over the world. From our despotic biotechnological throne, we can seek revenge on everyone who ever wronged us. What's that? A buzzing in your ear? I hope you're on OUR side."

All I can say is that I assume he is saying this with tongue in cheek. I really do.

CHAPTER 7

Parenting and the Rotten Corpse

I HAVE never understood why nature shows on animal families are always filled with images of doting monkeys nursing their infants, or diligent songbirds delivering a beak full of worms to the nest, when much more tender sacrifice takes place under leaf litter in the garden. If you want an ideal example of a good animal mother, for my money you can't do better than an earwig. Now there's a devoted parent for you. After they lay their eggs, earwig mothers stand protectively over the clutch, scrubbing them clean of fungus and other nasty contaminants and keeping predators at bay. Once the eggs hatch into minuscule copies of their parent, mama earwig goes resolutely out into the world to catch prey in the form of aphids and other tiny invertebrates for her brood. In some species, the female digests the food first and then regurgitates it to her begging offspring, as if offering a squalling infant a bottle. If the young earwigs signal their distress, she responds to the solicitation with eager defensiveness. Oh, and that business about them climbing onto people's heads and into their ears? Utter nonsense. According to entomologist James Costa, the name was probably originally *ear-wing,* after the resemblance of the insect's hind wing to a human ear (honestly, I don't see it, but that's an urban legend for you). How that got transmogrified into an auricular horror story is anyone's guess.

I happen to have an admittedly unpopular fondness for earwigs, but there are a lot of other good insect parents out there. The real champions, of course, are the social insects such as bees and ants, in which the mother goes into the hive or nest after a brief mating flight, never to emerge again. Day in and day out for months, sometimes years, the queen mother produces egg after glistening egg, like chocolates on an assembly line, forswearing any other activity. After the first batch, of course, the maternal care itself—the feeding, the cleaning, the guarding—is foisted off onto the queen's other offspring, but she is still at it, using the sperm she has saved from a dimly recalled frolic in the open air. She can never wait "until the children are grown" to do something for herself; her nest will never be an empty one, and her sacrifice is lifelong. I suppose it's true that not a lot of alternative careers are out there for older female ants and bees, but then many of the human mothers who dream that if only it weren't for the children they would be the CEO of a Fortune 500 company are not exactly being realistic either.

Comedian Milton Berle said, "If evolution really works, how come mothers only have two hands?" The answer is that at least some of the time, they don't—they have six. The social wasps, ants, and bees are highly specialized in their parenting, but many other kinds of insects, from true bugs to ladybird beetles, tend their young to varying degrees. Giant water bug females glue their eggs to the backs of the males, who carry the eggs around until they hatch. Assassin bug parents stand guard over their eggs and hatchlings. Ladybird beetles (a more entomologically correct name than *ladybugs,* since they lack the soda straw–like mouthparts of true bugs) produce infertile eggs alongside the more conventional ones, apparently solely to feed the voracious young beetle larvae when they hatch. And in some species of cockroaches, insects even more reviled than the earwigs, parents will remain together after mating, and females feed the young with special secretions, almost like milk. Sometimes females do all the parenting, sometimes males do, and sometimes both parents cooperate to raise their young. Finally, in most insects, no one cares for the young at all — the eggs are plopped unceremoniously in their place to fend for themselves after hatching.

All of this variation means that insects are an ideal place to look for insight into the evolution of family life. Why did parental care evolve in some groups and not others? Why does the amount of care, from a brief cleansing swipe over the eggs to putting the kids through college, differ so much among different kinds of animals? Why does mother do most of the work in many species, both parents in a few, and the father alone in fewer still?

Humans and other social mammals are of little help in answering these questions, because we always invest an enormous amount of energy into raising our offspring; we cannot compare situations where care is and is not given. Scientists often use birds to study the evolution of parenting, but these too are rather invariant in their behavior, at least when viewed next to insects. True, some birds, for example, ducks, are ready to waddle at hatching and need only judicious direction to the nearest body of water, while others, for example, robins, spend days or weeks of hard labor ferrying food to the gaping mouths in the nest. But that difference pales in comparison with the contrast between a butterfly who lays her eggs on a plant stem and flits away, and a pair of beetles who collaborate to prepare a ball of carrion for their larvae and then respond to the begging motions of their offspring by feeding them a liquefied meal.

People have dearly held opinions about parenting, of course, which is one reason it is an interesting and worthwhile behavior to understand; for example, we want to know if it is "natural" for fathers to take less of an interest in their daughters than their sons, or in being an attentive parent at all. But beyond questions such as these, understanding how the family evolved can help us understand the evolution of social life itself. At the core of society is the oldest bond between individuals that exists: the bond between a mother and her offspring. Once a female stays with her young, the stage is set for siblings to collude and squabble, for alliances to form and dissolve. That in turn makes more complex social interactions possible. In the wink of an eye — well, maybe in a few million years — presto, you have a society, with insurance companies and a movie industry and supermarket tabloids. Families become labor unions, political parties, and royal dynasties. And it all started with a female bug standing guard over her egg. Looking at insects can help us see how we got there. The true social insects, which include the ants, termites, and many of the bees and wasps, are a situation unto themselves and, as I discuss elsewhere, share unique genetic relationships that make the costs and benefits of their behaviors different than those in other organisms. I will therefore confine myself here to the "other" social groups, where only the first timid steps toward a complex network of interaction have been taken.

Why Should You Care?

So if most insects, and indeed most animals, get by with recklessly flinging eggs to the four winds, metaphorically speaking, why has more elaborate parental care evolved where it has? People often assume that human children require as much care as they do because of two characteristics of our species: high intelligence, sometimes seen as a reliance on learning rather than instinct, and being born at a comparatively early and helpless stage of development. Our intelligence, and the accompanying complexity of our lives, supposedly means that parents need to spend a great deal of time teaching their offspring the ins and outs of life in society. If we were simple little automatons, the conventional wisdom goes, we too could dump our babies into the world and let them fend for themselves. But faced with an unopened cereal box and a carton of milk, much less a gazelle or a yam, a child needs someone else to open, kill, or cook the meal, and hopefully that someone will make sure that eventually the child will be able to do the same thing unassisted.

The problem with that argument is obvious once you think about insects. Although they do learn more than had previously been believed, as I discussed earlier, not even I am going to champion their qualifications for Mensa. And yet many groups still show elaborate parental care. Furthermore, the species that have doting parents don't seem any more or less intelligent than those lacking them.

What about the idea that we are stuck with helping our children because they are born at such a nascent state? Our big brains mean that the female pelvis can't accommodate a baby born any later, according to this notion, so it is related to that "we are so smart we have to toddle around for years before we can manage to leave our mothers" idea. But here too the insects make that suggestion look dubious, since egg size doesn't have much to do with the size of the brain of the bug that is hatched from it. Earwig mothers, doting though they are, never labor to bring forth their six-legged progeny.

What is essential is the guiding principle behind the evolution of every trait, whether that trait is a behavior, such as offspring care, or the shape of a body part, such as the length of a tail. Doing it has to increase the bearer's fitness, the likelihood of passing on its genes, more than not doing it. When it comes to parental care, that means that even if tending the young takes away valuable time and energy that could theoretically be used to have more offspring, it's worth it if you leave more copies of your genes that way than you would if you deserted your young and went off to have more children. Having offspring is valuable, certainly. But it will win you the evolutionary jackpot only if those offspring survive. Rampant fecundity for its own sake goes unrewarded.

For insects, and maybe for the rest of us, the threat that made it worth giving up future offspring to focus on the current batch seems to be predation. It's a beetle-eat-beetle world out there, so to speak, and eggs are about as vulnerable a stage to be exposed to it as can be imagined. In a small, unassuming black bug called a burrower bug, females guard the mass of eggs they produce in the leaf litter on the forest floor. Once the young hatch, the mother feeds her babies with nutlets, products of mintlike plants that grow nearby. When Japanese researchers Taichi Nakahira and Shin-ichi Kudo removed a female from her eggs on the grounds of Hokkaido University, the eggs succumbed to predators or were attacked by fungus and virtually never survived to hatching. Female shield-backed bugs in the harsh deserts of Baja California crouch protectively over their tiny brood on the stems of croton plants, and if the mother is removed, her young are almost immediately eaten by ants and other insects. If they happen to fall to the desert floor without their mother's watchful presence, they shrivel and die. It looks like love. But it's just a smart investment.

The earwigs demonstrate this trade-off even more clearly, because unlike the burrower bugs, in some species of earwigs an individual female may or may not tend her young. Some broods get more care, others less. In a paper titled "Benefits and Costs of

Earwig Family Life" (with just a little bit of jazzing up, can't you see this as a sitcom?), Mathias Kölliker at the University of Basel in Switzerland pointed out that female earwigs sometimes lay two batches of eggs in a season, but if they invested in the work of protecting and provisioning the first set of young, they were less likely to produce a second. Even when they did lay another clutch, they did so later in the season than mothers who had abandoned the first clutch. That delay can be crucial when the cold weather threatens, which means that the benefit of nurturing the first batch of offspring has to be weighed against the cost of having the second batch die in the first storms of autumn.

Similarly, females of a kind of treehopper found in eastern North America will guard their eggs for varying lengths of time, but sometimes desert them right after they are laid. Andrew Zink painstakingly followed 370 female treehoppers over the course of an entire season, dabbing their bodies with colored paints so that he could identify individuals day after day. In terms of the benefits and costs of staying versus going, it turned out to be six of one and half a dozen of the other, or (in the case of the treehoppers) perhaps a hundred of one and five score of the other. Females that stayed longer had more eggs that hatched than the females that left after they laid their eggs, but the protecting females then had fewer and smaller broods later on. The bottom line is that taking care of children has to be selfish in evolutionary terms. A mother who died in the process of keeping one youngster alive under difficult circumstances would leave fewer copies of her genes than one who cut her losses but lived to reproduce again and again.

Sometimes it does pay to go for broke on the first batch of young, if life is so uncertain that survival to produce another brood is unlikely. In such instances, mothers sometimes offer up the ultimate sacrifice, as is detailed in a paper on a social spider by Ted

Evans and his colleagues in Australia, delightfully titled "Making a Meal of Mother." Spiders are not insects, of course, but I include this one because it is such a wonderful example of how evolution can produce apparently self-destructive behavior. As any *Charlotte's Web* aficionado knows, all spiders show some kind of maternal care, but in this one, the young spiderlings slowly suck blood from the leg joints of their mother while she is still alive. Gradually they consume more and more of her body, until, in Evans's words, "After several weeks, she is decrepit, unable to move, and the offspring eat her entirely." When the scientists weighed the young spiders, they had gained virtually exactly the amount their mother had lost. And being fatter at the outset meant you had more to offer; scrawnier mothers were consumed sooner than their more zaftig counterparts, which then meant that the young spiders were less likely to turn from their parent to an arguably even more unsavory occupation: eating each other. Having your children feed on your still-quivering flesh to keep them from cannibalizing their siblings sounds like something out of a parody of a Philip Roth novel. But for the spiders, it's sensible, since each brother or sister eaten is a genetic investment lost to the parent.

Under what circumstances do we expect these and other less dramatic efforts by parents to show up? In his classic book *Sociobiology: The New Synthesis,* Harvard biologist (and ant lover) E. O. Wilson lists the "prime movers" that make the evolution of parental care more likely. First is the threat of predators that I mentioned earlier. Other contributors include the kind of environment in which a species finds itself. If food is unpredictable or the climate harsh, offspring will benefit from a parent's buffering them from the vagaries of the world.

Doug Tallamy and Thomas Wood from the University of Delaware took this one step further and pointed out that in addition, the species that evolve parental care have to have several preconditions. They have to reproduce only during a narrow window of time and during a brief period of the year, simply to make parenting cost-effective. They have to survive long enough to provide enough care to be meaningful, a tall order for animals as short-lived as most insects. And because insect eggs don't last well unprotected, particularly during long cold periods, the adults can't spend much of the summer or other favorable time of year growing up themselves. Instead, they have to be mature, or nearly so, at the start of clement weather so that the young ones get the full benefit of their care over the nicer times of the year. And last of all, some behavior patterns that could be adapted to parental care have to be present already in the species; Tallamy has seen the same kind of aggressive display used by mother lace bugs in defense of their brood when the females are discouraging overly ardent males during the mating season.

These kinds of traits are sometimes called *preadaptations*, characteristics that happen to be present for some other reason and that can then be co-opted for a different use under different selective pressures. Natural selection can't use an empty palette; the raw material for parenting, or flight, or digesting a new food, or any other evolutionary innovation, has to be there already. Preadaptations aren't to be confused with evolutionary premonitions—you don't live in a climate with fleeting resources or bursts of cold because nature decided you should be a parent someday and thought those things would nudge you in that direction. They are more like the bits and pieces that happen to be lying around in your garage; if you want to build a bookshelf, and Home Depot doesn't exist, you are stuck using what you have available. So parenting used the lifestyles already out there and built upon them to make the patterns of care that we see today.

Your Turn, or I Got Up Last Time

Mothers are fonder than fathers of their children because they are more certain they are their own.

— ARISTOTLE

A MALE friend of mine says that when he used to take his son to day care, the other parents — all mothers — would stare at him like he was a child molester. Social progress of the last few decades notwithstanding, women in virtually all parts of the world do much more of the child care than men. It's also much more common for mothers than fathers to take care of the young in animals, insects included. Why is that?

As with many basic truths, Aristotle was onto something when it came to parenting. The concept he alludes to above is what behavioral biologists call *confidence of paternity.* It implies nothing about a conscious awareness of one's likelihood of having been cuckolded. A male bird, say, that fetches mouthful after mouthful of painstakingly collected caterpillars to the chicks whose beaks gape insistently in his nest will not pass on his genes if it turns out that the female mated with another male during her fertile period several days or weeks earlier. Even worse, he has wasted time he could have spent trying to mate with other females himself. The female, on the other hand, either laid those eggs or didn't, and she was there at the time (aside from exceptions such as brood parasitism, which I'll discuss later). In mammals, where the events of mating and birth are even more widely separated, the problem of knowing which babies are sired by which male is even worse. There, of course, only the female can supply the offspring with milk, but males can perform other fatherly acts, for example, protecting everyone from predators or getting food for the mother.

This disparity between the likely payoff to each sex that accrues from devoting oneself to one's offspring is often cited as the sole reason for males rarely being the sex that takes care of the young. It's evolutionary good sense to refuse to take on someone else's genetic investment, and males are generally thought to benefit more by competing for mates than by sticking with the offspring. But while certainty of paternity probably plays some role in the evolution of different reproductive behaviors, it is now emerging that it can't be the whole explanation. It's all very well and good to say that a male "should" go off and seek other females to mate with rather than stick around and care for the ones his current mate produces. But what are his chances of succeeding? Everyone likes to think he could have been a contender, but in reality, it's tough out there. Females may be scarce, they may be unwilling to mate with every philandering male that passes by, and the risk of being eaten by a predator before one is found may be high. Tip the balance of any one of these variables, and the parental care patter can skew toward mom, dad, or both. And insects are perfect test cases for ideas about "innate" parenting roles, because once the eggs are produced, either sex can easily protect or bring food to the babies, unlike mammals, where females have all the milk-producing apparatus.

Single fathers are rare among insects, as they are elsewhere, but when they occur they do a bang-up job. Giant water bugs are true bugs, meaning they have strawlike mouthparts that they use to suck up their food. Some bugs, for example, aphids, merely sip sap, but the giant water bugs are fierce predators that ambush prey such as other invertebrates, fish, salamanders, and frogs. Once they grasp their victim, the bugs inject enzymes that liquefy the contents so that the interior can be extracted. About the size of an almond with the shell on, the bugs patrol lakes and streams around the world.

When it is time to breed, males attract females by suggestively rippling the water's surface. Unlike most insects, after mating, instead of the female taking off with her store of sperm, giant water bug females from most species proceed to lay eggs on the back of the male, where they sit in neat pearly rows until it is time for them to hatch. Although the conspicuous bugs were noticed by early naturalists, the individuals bearing the eggs were assumed to be females. Even when the sex of the brooding individuals was known, scientists as late as 1935 declared that the females must be forcing the males to carry the eggs. The father, however, always mates with the female before allowing her to deposit her eggs, and the pair may go through several rounds of mating followed by egg laying, apparently at the male's behest. This probably ensures that at least a majority of the eggs carried are indeed the male's own. He then solicitously ensures the eggs are supplied with oxygen by periodically raising his back above the surface. Although more than one female can deposit her eggs on a given male, the space on his back may become limited, and the males do what they can to position females so that they lay eggs to fill in any gaps in the row. Carrying around the eggs is no easy task; their combined weight can be twice as heavy as the male himself. In one group of giant water bugs, the females lay egg masses that are attached to vegetation at the water's edge, and the males then guard the eggs while the female departs, perhaps to lay another batch fertilized and protected by another male. The male stays to guard the young bugs even after they have hatched, preventing both predation and cannibalism from occurring; a photograph of one such species shows the tiny striped nymphs clinging to a plant stem, as if to a pool toy, while their father bobs beneficently nearby.

Biparental care, in which both parents cooperate to take care of the offspring, is at least as rare as male-only care among animals. It is seen in a few insects, however, including some that have the unfortunate attribute of being able to clear a room better than any other species. When I was doing my doctoral research at the University of Michigan Biological Station, the state-of-the-art Alfred H. Stockard Lakeside Laboratory had recently been completed. It was a lovely building, with excellent facilities for studying all kinds of local flora and fauna, from algae to woodpeckers. I was lucky enough to have a room to myself on the second floor, where I housed my crickets and recording equipment. Most days I worked happily at the microscope or with my experimental subjects, chatting with the other students and faculty, and generally having a grand time. But sometimes one of us would see David Sloan Wilson or one of his helpers walking toward the building with a white bucket, and we all knew to scatter. A multitalented evolutionary biologist, at the time Wilson was working on burying beetles, and the stench was enough to make paint peel.

As their name suggests, burying beetles locate recently dead animals by smell. If a male arrives at a carcass first, he sends out a chemical signal to attract a female. The pair then prepares the carcass as a nursery by stripping off any fur or feathers and shaping it into a ball. They cover their prize with specialized secretions that deter mold from growing (though this does nothing to deodorize the body, as we discovered to our regret). They then dig in the soil underneath the carcass, allowing it to sink into the forest floor. Once the carrion ball has been safely sequestered underground, the pair mates. The female then lays her eggs in the soil surrounding the carcass, and when the eggs hatch, the larvae beg for food from their parents by turning their heads toward mom or dad and waving their tiny arms. They are fed with either bits of carrion carved off the carcass like slices from a roast, or regurgitated meat that has been partially digested by a parent.

Although this lifestyle has much to recommend it (a steady and nourishing food supply, protection from predators once the carcass is buried), burying beetles face a daunting problem: finding a suitable dead animal, and once it is found, defending it against rivals. Even avid hikers and wilderness lovers rarely come across a dead animal at all, much less one that is both small enough to handle and still fresh enough to provide the right environment for raising young. The beetles have extraordinarily sensitive odor detectors on their antennae, and can sniff out a carcass from miles away, but even so it is an extremely difficult undertaking. Recently acquired carcasses are therefore highly sought after, and if a burying beetle comes across a carcass that has already been colonized by another individual, pitched battles may result. Both parents participate in defense of their property, and if the male happens to be away from the carcass when a male intruder threatens, the invading male will kill the offspring and mate with the female. Female intruders are not as successful in taking over a carcass. If both parents are present, they can usually fend off the invaders and hang onto their prize. It is thought that this advantage of pair defense is what led to the evolution of parental care by both the mother and the father in this insect group. If the carcass is large enough, multiple females may stay and lay their eggs, although one of them is generally dominant over the others.

Perhaps because of this difficulty in finding carcasses, one species of burying beetle has abandoned the role of "nature's undertakers," as one website refers to them, entirely. The substitute for dead birds and mammals that they chose, however, has to give one pause. They didn't switch to eating live insects, say, or any kind of vegetable matter. Instead, they are found in snake nests, eating the eggs and feeding them to the offspring. The beetles do not bury the eggs, since the snake takes care of that herself, and then leaves, so that the beetles do not have to face an irate mother snake. It seems to me that this was not the most sensible choice to have made — quick, which would you find easier to locate in a forest, a dead mouse or a snake egg? — but presumably it allows the beetles to exploit a resource that is less likely to be taken over by competitors.

What Happy Families?

I HAVE a *Six Chix* cartoon depicting two generic-looking insects sitting in armchairs; one is obviously bloated and looks a bit guilty, and is being reassured by the other, "It's natural to eat your young, Marilyn . . . especially when they start running around the house like that — sometimes you just lose it."

Insect mothers do indeed sometimes eat their young, but it's not annoyance that elicits the behavior. Infanticide and subsequent cannibalism are yet another manifestation of the rule that parenting is worthwhile only if it furthers the parent's interests. If investing in offspring now means losing out on future opportunities to reproduce, natural selection will not favor it. But a problem can arise if the world looked a certain way when a mother first produced her offspring and then changed once they were a little older. Because insects live their lives so quickly, often relying on transitory sources of food and shelter, and because they can replace a batch of eggs with relatively little trouble, simply offing one generation of young and starting from scratch is a more reasonable proposition for them than for at least some vertebrates, who need a lot of time to gin up another generation. This makes them excellent subjects for studying the circumstances under which infanticide is favored.

Say a mother beetle lays a batch of eggs when the environment is benign and food abundant. She will benefit most by depleting her fat stores and turning them into eggs, since she can replenish those stores with the food around her. The eggs will survive best if she guards them on the plant, since predators would eat them if she were not around. But unexpectedly, the food supply dries up — maybe the gardener stops watering her food plant, or a cold snap makes it hard for her to move around and eat. What should she do? If she could read, she would be advised to take a look at an insightful paper by Hope Klug and Michael Bonsall pithily titled, "When to Care for, Abandon, or Eat Your Offspring." In it, the scientists outline the circumstances that favor each option. Cannibalism of offspring is particularly likely to evolve if parents can be selective about which young they eat, focusing on the lower-quality ones. Eating eggs is also expected to be common if doing so increases the parent's reproductive rate later on.

These principles are nicely illustrated in a group of insects with the delightful name of assassin bugs. These often brightly colored bugs ambush prey from their perches on vegetation, and in many species one or another parent guards the eggs. One African species is unusual because the father, rather than the mother, is the caregiver. A male will guard the eggs laid by several females, so he does not lose future mating opportunities by running the bug equivalent of a day care center. Males frequently eat a portion of the eggs but tend to focus on those at the edges of the cluster. Interestingly, Lisa Thomas and Andrea Manica from the University of Cambridge in the United Kingdom found that those peripheral eggs were the most likely to have been parasitized by a tiny wasp and, hence, were going to yield baby wasps rather than miniature assassins. It doesn't appear that the males can tell which eggs are parasitized, because in the wasp-free laboratory they still eat eggs from the same position in the cluster as they do in the wild. Instead, natural selection presumably favored fathers that removed fewer of their future offspring from the gene pool. The eggs are an important food source; cannibalistic males didn't lose any weight while they were guarding, even though they couldn't go out and hunt.

Infanticide and subsequent consumption of young was frequently observed in laboratory animals such as rats, and for many years the behavior was interpreted as abnormal and pathological, an artifact of captivity. Its documented occurrence in insects somehow didn't seem relevant to people, perhaps because we don't automatically see ourselves mirrored in their behavior. But then it began to be seen in wild animals such as lions as well, and now it is clear that at least some of the time it is probably adaptive in nature, because rearing young when life is harsh, or at the expense of the parent's well-being, may be too big a gamble for it to be continued. (Several kinds of animals, including lions, also commit infanticide without cannibalism, and of others' rather than their own offspring, for different but equally adaptive reasons.) If the going gets tough, the tough—and the smart—stop taking care of their children.

Some insects even go so far as to produce infertile eggs, called trophic eggs, that they either eat themselves or use as extra provisions for the young that do hatch. Although in some cases this behavior may be opportunistic, with some eggs not developing because they are defective, in others the trophic eggs seem to have evolved as a food source. Ladybird beetles are particularly well known for the production of such extra meals and will lay more trophic eggs when prey are scarce, fewer if they are well fed. Starved female ladybird beetles may lay an egg and then immediately turn around and eat it, a much tidier solution to hunger than, say, gnawing off a limb. The trophic eggs often look different from those that develop normally. Putting extra provisions into additional eggs rather than simply making larger eggs that hatch into more robust offspring with the surplus may have evolved because mothers cannot manufacture larger eggs with more yolk reserves than they already do. Scientists suspect that trophic eggs may be cheaper to produce than the usual variety, although the details of this cost difference are still not clear. The trophic eggs can also deter earlier-hatching babies from eating the eggs of their tardier siblings, which means that the mother gets more offspring surviving to maturity.

In addition to being excellent subjects for examining cannibalism, insects are perfect for exploring another stark reality of family life: parent-offspring conflict. In extremely influential work published in 1974, Bob Trivers, the same biologist who worked out some of the niceties of sex ratio theory discussed in an earlier chapter, pointed out that while parents and children have half their genes in common, they don't necessarily both benefit from the same things. Imagine that a mother beetle has a brood of twelve offspring. All else being equal, natural selection will favor her giving equal amounts of food to each of her children, because they are each equally related to her and equally likely to pass on her genes. But from an individual offspring's perspective, getting more care for itself at the expense of its siblings will also be favored, since it is 100 percent related to itself and only 50 percent related to its siblings. Thus there is a difference of opinion, evolutionarily speaking, in where the attention should go. Trivers called that difference of opinion parent-offspring conflict, and it is now thought to occur in a wide variety of animals and even plants. Trivers's theory explains many seemingly paradoxical family behaviors, including the grimmest: infanticide, discussed above, and its cousin, siblicide, the killing of one's siblings.

At some level, everyone with siblings understands the urge to murder them. Parent-offspring conflict theory suggests that such desires are not necessarily maladaptive. For the ladybirds, cannibalism is a particularly potent risk because the beetles' bright colors indicate the presence of toxins in the body, and a noxious taste to go with them. Your average bird learns to avoid ladybirds, but the beetles themselves are not affected by the chemicals, and so a ladybird is its own worst enemy. Eating just a single egg, whether trophic or not, significantly boosts the growth rate of a ladybird larva, putting a premium on hatching early. In a study of European ladybird beetles, less than half of noncannibals made it to adulthood, but more than 80 percent of the larvae that had eaten a sibling matured successfully.

Earlier laid means earlier out of the egg, with a chance to prey on one's less precocious siblings. But the day that an egg hatches relative to its nest mates is largely under the control of the mother, since she, after all, is the one that puts them all there. This fact brings up an even more sinister aspect of siblicide: frequently the parent encourages it, or at least does nothing to keep it from happening.

Why should parents tolerate such shocking behavior? Once again, think of the offspring not as individuals in their own right, but as units that pass on the parents' genes. If a mother lays many eggs, and they all hatch, but then there isn't enough food for them all, they all starve, and she has lost, big time. If she lays only a few, and it turns out that there would have been enough food for more, she has still lost, albeit not as much. But if she goes ahead and overproduces, and then lets sibling rivalry take its course, she can achieve the golden mean, since exactly as many offspring as the available resources can support will have survived. So a mother that lays her eggs over several days, creating a situation in which they hatch at different times, may be hedging her bets; if food is abundant, everyone gets enough after hatching and all is well. But if not, the earlier-hatching individuals can eat their siblings and ensure that at least some of the offspring make it to adulthood.

Even if they do not actually consume their brothers and sisters, competing for limited food supplied by a parent, and shoving a less insistent sibling away, can achieve the same effect: more food for the one that shouts the loudest. Encouraging the competition, or at least turning a blind eye to it, will benefit the parent more than trying to break up the squabbles and ensure that food is divided equally among the members of the brood. Actual siblicide, rather than garden-variety making one's brothers' and sisters' lives miserable, is expected only under extreme circumstances, since of course the siblings have half their genes in common, and so eliminating them entirely has its costs as well.

Here too is where the parental and offspring interests diverge. No offspring is expected to willingly sacrifice itself, and indeed each one should attempt to get more food and attention for itself than its siblings. This greed is even expected to extend to the hypothetical future siblings the parent could produce; from an offspring's perspective, getting as much as it can right now will benefit it more than allowing its parent to keep some energy in reserve to invest in future offspring. But the parent should only give as much as necessary for the offspring to become independent and able to fend for itself, because the parent will be served best by saving some resources to produce offspring later. The resulting conflict, according to Trivers, leads to weaning tantrums in mammals and many other kinds of behavior in which young animals compete fiercely with their siblings and try to get more than their parent is willing to offer. Social insects, because they share different proportions of their genes with their siblings than their parents or offspring, are particularly prone to these kinds of disagreements.

If all of this sounds uneasily familiar from your own life, or perhaps the lives of some of your friends, you are not alone. One of the foremost theorists in the field of parent-offspring conflict, H. Charles J. Godfray, notes, "There are clear dangers in overinterpreting such behaviors (especially as they are observed through the prism of one's own family experiences)." This is where the insects come in handy; it is harder to anthropomorphize gleaming white grubs wriggling near a mouse corpse than, well, virtually any other baby animal I can think of. Although much of the early research on the evolution of parent-offspring relationships in nonhumans used birds as subjects, scientists are increasingly realizing that the begging behavior of burying beetle young is no different than that of a robin squawking in its nest, and it is much easier to use in experiments.

Not My Problem

RATHER than deal with quarreling offspring or hungry mouths, some insects whose young require care after the eggs are laid have abdicated at least some of that care entirely. We've long known that cuckoos, cowbirds, and a few other bird species are brood parasites, which means that females lay their eggs in the nest of another species, the host. Some ants do much the same thing by using a different species of ant to rear their young, either by capturing eggs of the foreigners and bringing them back to the nest or by killing a queen and replacing her with one of their own kind.

More recently, a more subtle but no less effective means of getting someone else to do the work of child rearing has been recognized in both birds and insects. Rather inelegantly but descriptively called *egg dumping,* it means exactly that: depositing eggs into the nest of another member of the same species. Often a female that practices egg dumping still cares for some of her own eggs, but the farmed-out offspring serve as a kind of bonus rainy day account, allowing her to literally not put all her eggs in the same potentially vulnerable basket. In other cases, skipping out on maternal care means that a mother can keep churning out batch after batch of eggs at a rate that a mother spending time and energy on demanding offspring could not equal.

Lace bugs, delicate insects with filigreed wings that live on a variety of garden plants, face a trade-off between protecting their young from predators and losing future reproductive opportunities by doing so. Doug Tallamy has studied maternal care in several kinds of lace bugs, and found that they will keep laying eggs in the egg masses of other females and abandon them to the care of the female that first started the egg mass. They keep doing so until they cannot find another suitable host, at which time they proceed to guard their own eggs and offspring, presumably along with those that other females have foisted onto them.

The burying beetles will also engage in a little stealth egg laying if their carcass has been taken over by another pair. The defeated female will stay near the dead animal and sneak into her former nursery to feed on the carcass herself and surreptitiously lay some eggs. Sometimes, if the carcass is large enough, she is even tolerated by the female in possession of the corpse, and both remain to rear their young in the same underground chamber.

As with infanticide, egg dumping used to be dismissed by biologists as an aberrant behavior that occurred because the female was too stupid to figure out how to breed properly; an early researcher who discovered egg dumping in ducks rather pejoratively called it "careless laying," and "degenerative." Presumably the idea that a mother would callously abandon her own offspring hit a little too close to home, as Godfray suggests above. Perhaps because no one expects insects to be smart in the first place, or identifies with a bug on a leaf, these biases have not gotten in the way of scientists developing hypotheses about the evolution of egg dumping in insects, another illustration of how using insects rather than vertebrates as models can make it easier to understand behavior.

One's immediate thought is that the dumpee, or host, is a sad patsy here, but unlike the case for brood parasites such as cowbirds or cuckoos, where the host's own offspring virtually always suffer as a consequence of the interloper's demands, egg dumping can actually benefit everyone concerned. Predators often do not gulp down the entire egg mass or group of babies. Instead, they usually nibble off a few eggs here and there, or they pluck the most vulnerable larva from a cluster. In such cases, it pays to be part of a teeming horde, because one's chances of being the unlucky victim go down the more options the predator has; if a wasp or spider snatches one egg, and ten are present, you have a one in ten chance of being eaten. But if one hundred eggs are in the mass, your chances go down to one in a hundred, much better odds. At the same time, guarding a mass of a hundred eggs isn't much more work, if any, than guarding ten. In a North American treehopper that lives on goldenrod leaves, the hatching success of broods that were supplemented with dumped eggs was 25 percent higher than that of broods reared with only their own brothers and sisters.

Given this happy everyone-wins scenario, why isn't egg dumping even more common than it already is? Tallamy suggests that the opportunity to dump eggs may be constrained by the size of the nest (if eggs have to be placed in a particular spot, for example, a stem, at some point there is no room left), the physiological capability of the female to keep producing eggs, or the synchrony among females in their reproductive stage. It does no good if you have eggs to offload if all the other nearby females are already half-

bee has brought back for her own offspring. Finally, they emerge as adults, to start the whole improbable cycle again.

Where did this bizarre life cycle come from, and why do the beetles rather than other—perhaps many other—kinds of animals exhibit it? We don't know for certain of course, but the sheer number of insect species may have provided a larger canvas on which to paint different pictures. Many of them probably died out before they became established, but a few, like this one, remained.

Although this scenario brings up many interesting issues, from the standpoint of family relationships it is like one of those conundrums of which the philosophers are fond. A male bee can hold only so many larvae, a fraction of the aggregation. But all of the members of the group have to cooperate to constitute a convincing mimic of a female bee. The remaining larvae can try again, to be sure, but as their numbers decrease, their portrayal of the bee loses its verisimilitude. This leads to what must be an increasingly uneasy alliance, as the beetles' only hope for survival arrives and the larvae must jostle for a chance to leap onboard. When should a larva stop working harmoniously, break rank with its brothers and sisters, and act solely in its own interests? Hollywood, take note. In my opinion, a movie based on this kind of drama has a lot more appeal than yet another tired take on the dysfunctional family reunion at Thanksgiving.

CHAPTER 8

Pirates at the Picnic

Today it is accepted as proven that the ant is incontestably one of the noblest, most courageous, most charitable, most devoted, most generous and most altruistic creatures on earth.

— MAURICE MAETERLINCK, 1930

If ants had nuclear weapons, they would probably end the world in a week.

— BERT HÖLLDOBLER AND E. O. WILSON, 1994

ANTS may inspire more emotional reactions than any other insect, reactions that go far beyond the revulsion of finding a cockroach scurrying across the kitchen counter or pleasure at seeing a butterfly light on a flower. As the two quotations above attest, ants can be paragons of harmony and virtue, or symbols of bloodthirsty violence. Honeybees come close to ants in serving as reflections of our own society, but we see bees singly, flying from blossom to blossom, rather than en masse, and the workings of the hive are not visible to most of us. Ants, however, stream across our driveways in glistening black ribbons. They seethe through our cereal boxes and bear crumbs triumphantly along edge of the shelf and out the door. With a few moments of casual observation, it's possible to see ants carrying their young from place to place, whereas no one other than beekeepers (and entomologists) ever sees much in the way of bee family life. And they walk, rather than fly, making them a little easier, perhaps, to identify with.

Like many of the other social insects, ants seem to share food unhesitatingly, and they work tirelessly for their colony, as Maeterlinck notes above. Maeterlinck, a Belgian playwright and poet who won the 1911 Nobel Prize for literature, was particularly taken by the ants' practice of passing droplets of food from one individual to another, called *trophyllaxis,* a behavior not seen in most nonsocial insects. For reasons that are not altogether clear, at least to me, he seemed to think that this behavior was intensely pleasurable for the ants, somehow compensating the workers for their lack of sexual activity by a near-orgasmic sensation when the food was transferred.

Solomon, of course, enjoined us to "Go to the ant, thou sluggard; consider her ways, and be wise." As historian Charlotte Sleigh writes in her entertaining book *Ant,* "The ant's supposed virtues of industry, prudence and mutual aid were extolled by a great number of people." The diligent ants that labor and save for the winter provide a smug contrast to the grasshopper that fritters the time away in a host of moralistic fables. A fair share of people probably sympathize with the indolent pleasure-seeking half of the story more than with the ants themselves, but the object lesson is clear: hard work is virtuous and will be rewarded. The Victorians seemed especially partial to the idealization of the ants' nobility and stressed the idyllic domestic activity supposedly taking place in the ant nest.

But ants also have a dark side that is obvious to even a casual observer. Naturalists since ancient times noted the apparent wars that raged between ants of different colors, with battles that went on for hours. Army ants are so named for their rampaging behavior. And as Sleigh points out, "The commonly known fact that ants engage in warfare has given them a particular edginess in times of human conflict." And a handful of species of ants exhibit a behavior that is strikingly similar to slavery in humans: one kind of ant will make raids on a colony of another species and steal its young workers, to be reared in the nest of the invaders and put to work for the rest of their lives. Charles Darwin described part of such a raid in *The Origin of Species,* musing on "the wonderful instincts of making slaves." According to Bertrand Russell, "Ants and savages put strangers to death," although plenty of familial slaughter takes place as well. So-called killer bees are a close second, with plenty of media hype about enraged swarms pursuing hapless passersby. The pursuit is obvious (though the actual numbers of people attacked and injured is often exaggerated), and people stand at the ready to attribute rage and bloodlust to the pursuers.

Hostility has also been linked with insects in novel ways. A now-defunct band from Houston, Texas, was called "Insect Warfare." Its album *World Extermination* is being re-released by the deliciously named Earache Records, with apocalyptic cover art showing giant cockroaches, or possibly crickets (I am personally offended by this), fleeing a skeleton looming over a decaying cityscape. Humans are nowhere in sight.

So which is it? Do ant wars and slave-taking raids mean that these, and perhaps the other social insects, are particularly aggressive, and hence that warmongering is natural in animals? Does the devotion and self-sacrifice so approvingly cited by Maeterlinck prevail? A closer look reveals that the real villainy takes place much more surreptitiously, and while less full of carnage, it is far more deadly.

An Army of Savage Lace

WHEN I was a child I went through a phase in which I told people I wanted to be a myrmecologist when I grew up. Although I did indeed spend time watching the ants in our backyard, along with the other insects, I was probably driven more by smug delight at knowing that the word means someone who studies ants than by any actual career motivation. Be that as it may, when we had an assignment in third or fourth grade to read a book and report on it to the rest of the class, I chose a book on ants, and happily launched into a litany of their amazing behaviors. Ants, I proclaimed, made gardens of fungus that they harvested for food. They stored honeydew in their own massively swollen abdomens and fed it to the other workers, droplet by droplet. Not only that, I cheerfully told my classmates, who were by that point probably unnerved if they were not simply bored, but army ants could swarm through entire jungle villages, consuming every living thing they encountered by tearing it to pieces. Cows, pigs, chickens, and people, all were subject to the advancing hordes with their bladed jaws. If one were caught unawares by the oncoming troops, the only recourse was to set one's bedposts in saucers of kerosene, get under the covers, and pray the ants didn't find a way to drop down onto the bed from the ceiling. I was slightly hampered in my explanation of this dire state of affairs by my uncertainty of exactly what kerosene was, but I was sure that if I lived in an area frequented by army ants, I would be able to procure some.

Here my teacher intervened. Surely, she said gently, you are exaggerating. Ants couldn't possibly be that destructive. Perhaps they attacked the animals near the area, or got into a hut or two, but this scale of devastation and carnage seemed a bit much for such tiny creatures.

I dug in my heels. No, I insisted, the book had said (and hence I unswervingly believed) that the ants could tear apart a person in minutes. It wasn't just the odd hen or two, it was An Entire Village. I honestly don't remember exactly how or if this disagreement was resolved, or if my grade on the book report was reduced due to my teacher's suspicion of hyperbole, but I remain convinced that people don't fully appreciate the wonders of ants, perhaps because they refuse to believe the extraordinary things ants can do.

Army ants in particular inspire superlatives. They were described in detail in the late nineteenth and early twentieth centuries by legendary scientists such as William Morton Wheeler and Theodore Schneirla; the latter published a paper in 1934 titled "Raiding and Other Outstanding Phenomena in the Behavior of Army Ants," in the prestigious *Proceedings of the National Academy of Sciences,* which was, interestingly, placed into the "Psychology" section of the journal, as if it had more relevance to the workings of the mind than to zoology. Wheeler referred to army ants as "filled with an insatiable carnivorous appetite and a longing for perennial migrations, accompanied by a motley host of weird myrmecophilous camp-followers," and Hölldobler and Wilson call them "the unstoppable, superorganismic grim reapers of the tropical forest."

Army ants occur in several parts of the world, including the southern and western United States, but have been best studied in the New World tropics. They lack a fixed nest site, instead creating bivouacs, football-shaped masses that can be nearly a yard wide with anywhere from ten thousand to seventy thousand workers surrounding the queen and larvae, depending on the species.

Hölldobler and Wilson estimate that the bivouac contains "a kilogram of ant flesh." The ants link their limbs and jaws to form their shelter, making a kind of savage yet delicate lacework of individuals that supports layers upon layers of brownish bodies. Just after dawn breaks, the bivouac seethes and breaks apart, sending out lines of ants in many directions.

The lines include army ant workers of several shapes and sizes, all female, of course, despite my students' disbelief. The small and medium-sized individuals lay down an odor trail as they walk down the middle of the track, while the larger soldier caste ants, with their scimitar-shaped jaws, lumber alongside. Workers are about the size of many North American ants, perhaps as long as a grain of rice, but the soldiers are three times their size, about as long as a kidney bean. The streams of advancing ants have no leader; individuals hustle back and forth at the edges of the swarm, altering direction as they encounter prey.

As my childhood reading experience suggested, army ants and their relatives the African driver ants are merciless when they encounter an animal in their path. People and other vertebrates such as birds or squirrels, however, are usually able to evade the advancing columns unless they are injured or otherwise prevented from moving out of the way, which vindicates at least some of my teacher's skepticism, though the ants certainly could overpower an immobile human being. Insects, spiders, and other invertebrates usually cannot escape so easily and are surrounded by the eager jaws of the horde. Hundreds of ants sink their mandibles into the prey, their grip so strong that if the ant is torn away, its jaws remain imbedded in the flesh of its victim. Anecdotally, at least, this powerful grasp has led to their use as sutures by the Masai of Africa, who induce the ants to latch onto either side of a wound with their jaws, holding it shut even after the body of the ant is discarded.

(I was once asked how the ant jaws are removed once they have served their purpose, and I don't know the answer, save that the process might put that "ouch" at the tug of a Band-Aid to shame.) Small animals are borne away intact, while larger victims such as tarantulas or grasshoppers, or the occasional unlucky mouse or even deer, are efficiently butchered and carried off in chunks that can be managed by one of the medium-sized workers.

The intimidating-looking jaws of the soldiers, like many weapons, are not actually useful for practical tasks, and so all of the work of hauling food back and forth is done by the more modestly equipped smaller workers. Sometimes a group of such ants collaborates to transport a larger prey item, balancing it expertly among themselves so that the load can be carried by the minimum number of individuals needed. Schneirla reported that the entire operation is accompanied by the sound of thousands of tiny exoskeletons tapping against the dry leaves of the forest floor, a sound that according to Hölldobler and Wilson "beats on the ears of an observer until it acquires a distinctive meaning almost as the collective death rattle of the countless victims."

The naturalist and author William Beebe once observed a bivouac of army ants that had taken temporary residence in the outhouse near his laboratory in Guyana. Transfixed by the sight, he determined to observe the insects as they set up their encampment. He first noted the odor of the group, which was "sometimes subtle, again wafted in strong successive waves. It was musty, like something sweet which had begun to mold; not unpleasant, but very difficult to describe." He was deterred from further rumination by "a dozen ants [that] had lost no time in ascending my shoes, and, as if at a preconcerted signal, all simultaneously sank their jaws into my person." Beebe proceeded to take a chair into the outhouse and use the traditional technique of placing each of its legs into a can of disinfectant; he then rushed over to the chair, hung a bag of equipment over the back, and pulled his legs onto the seat. "Close to my face were the lines ascending and descending, while just above me were hundreds of thousands, a bushel-basket of army ants, with only the strength of their thread-like legs as suspension cables. It took some time to get used to my environment, and from first to last I was never wholly relaxed, or quite unconscious of what would happen if a chair-leg broke, or a bamboo fell across the outhouse."

This rhythm of activity, with the ants alternating between going out on raids and forming bivouacs, continues for months. In the Central American army ants that Schneirla studied, the ants sometimes will form a new bivouac every evening and sometimes settle in their self-manufactured housing for a few weeks at a time. Because army ants have no permanent nest site, they do not reproduce as many other ants do, with the release of winged males and females that mate in flying swarms before the newly inseminated queens found new colonies. Instead, at least in the species of army ants that have been the best studied, although both fertile males and females are produced at a certain time of year, only the males can fly. They attempt to join the bivouacs of another colony. At the next raiding period, a group of workers stays with the old queen and moves to a new bivouac, while one of the virgin queens is surrounded by another set of workers and travels to a different site, where she mates with one of the males that had flown into the colony. In some species of army ants the young queen mates with several males in succession, in others with only one. The male ants, like the drones of honeybees and many other social insects, die soon after mating, assuming they get a chance to mate at all. The rest of the handful of reproductive females are abandoned to the company of a small group of workers, but they do not hunt for food, and eventually all die, leaving their mother and sister to carry on in their place.

Army ant queens themselves have dramatically episodic reproductive lives; instead of monotonously laying egg after egg, day after day, for their entire adult life span, the queen's ovaries will develop only while the colony is in their more long-term bivouac. At that point, rapidly making up for lost time, her abdomen distends and she lays up to three hundred thousand eggs in one fell swoop. When the workers that are the product of her labor appear, they seem to perk up the energy levels in the group, and after a while the entire colony starts the migratory phase again, as the queen's labor subsides and she is shielded from harm as the columns of scissor-jawed daughters resume their activity.

The term *army ant* is not a truly scientific designation; it is used to describe those species of ants that exhibit both incessant migration of the entire colony and coordinated group hunting, including the raids by large numbers of individuals and the carrying of prey back to the nest. Sometimes terms such as *legionary* or *driver* ants are used, but virtually everyone who has written about them describes the ants' behavior in the most aggressive terms. While Hölldobler and Wilson, in their monumental tome *The Ants*, concede, "Yet driver ants are not really the terror of the jungle as popularly conceived," they devote much of their chapter on army ants to the same bloodthirsty details relished by Beebe and other naturalists. Even Maeterlinck, with his enthusiasm for the social virtues exhibited by ants in general and for their food-sharing proclivities in particular, sorrowfully notes, "Even in the ants this universal charity, this perpetual communion, does not prevent wars: though the wars of the ants are less frequent and less cruel than is generally believed." He also saw—or thought he did—a perfect mirror of human foibles: "Every kind of warfare known to ourselves will be found in the world of the ants; open warfare, overwhelming assaults, levies en masse, wars of ambush and surprise and surreptitious infiltration, implacable wars of extermination, incoherent and nerveless campaigns, sieges and investments as wisely ordered as our own, magnificent defenses, furious assaults, desperate sorties, bewildered retreats, strategic withdrawals, and sometimes, though very rarely, brawls between allies, and so forth."

In the midst of this veritable battalion of military metaphors, it might be worth stepping back and considering the actual goal of the army ants themselves. Those forms of warfare Maeterlinck exhaustively details might be better replaced by a far humbler list: going to the grocery store, harvesting vegetables in the garden, or hooking fish in a river. The ants are an army without an enemy. They are predators, and predation is not waging war, it is acquiring food. We seem to like linking hunting live prey with being aggressive, and we seem to especially like linking it to manliness. Predatory animals such as hawks or lions are often depicted as being exceptionally fierce, and so we transfer that to the ants, struggling with the grasshoppers that are elephant sized to them. But the truth is that hunting is a more widespread and less glamorous profession than it is sometimes made out to be. We tend to think of predators as animals that subdue large, warm-blooded prey, usually after a heroic struggle, but there is no a priori reason for us to dismiss animals that catch and kill more modest fare, for example, the ladybug gleaning aphids from a rosebush, or a chickadee nipping an inchworm off of a leaf. Some biologists refer to any food item that comes in a discrete chunk, as opposed to the unending sea of grass in a field, as "prey," and talk about animals such as the seed-eating kangaroo rats as "seed predators." Even if that is going a bit too far for some, is it any less savage to bite a worm than a weasel? Why does a hawk swooping down on a mouse seem more aggressive than a songbird snapping its bill against the hard shell of a beetle?

It's true that hunting, for both humans and other animals, can be risky, and facing up to prey that is bigger than you are and that has sharp teeth or claws can take courage. And in some cultures hunting, because it requires that bravery, is used as a test of manhood. But none of this applies to the ants, not least because, of course, all of the workers—even the ones with the big, bladed jaws—are female and won't get any more kudos from the colony no matter how many tarantulas or pythons they bring down. That they all eat meat doesn't make them any more vicious than the more peaceably named harvester ants that lug heavy seeds back to their nests.

Maybe the emphasis on warfare and aggression in army ants is an effort to counter that idealization of the ants' social harmony that used to be so prevalent. After all, Solomon wanted us as sluggards to look to the ant for inspiration to hard work; he didn't ask us as wimps to look to her for inspiration to violence. Sleigh discusses a book on "natural history and animal morals" published in 1851 by the Society for the Promotion of Christian Knowledge, which is still active. In it, ants are held up as models of prudence and industry from which humans are exhorted to learn. Ant and bee societies were embraced by the natural theologians of the nineteenth century, and their cooperative altruism was used as a model for social organizations that supported the poor. Later, socialists claimed that ants were exemplars of comradely sharing, though they must not have looked too closely at that turgid-bellied queen.

As my childhood flirtation with a career in myrmecology suggests, I am hardly immune to the drama of the ants' activities. But all of this symbolism, and the focus on aggression, runs the risk of seeing the ants as miniature soldiers instead of the skillful predators they are. What's really interesting about army ants isn't whether they show "incoherent and nerveless campaigns" as opposed to "sieges and investments as wisely ordered as our own." It's how and why that one virgin queen is singled out and sequestered from the rest in the colony. (Is she older? Younger? Or does one live and the others die at random? We still don't know.) It's how the colony's internal clock tells it when to migrate and when to make camp, given that research has shown the ants are not simply driven by hunger. Day length may give a proximate cue, while colony size also plays a role. It's how those massive colonies in Africa and South America differ from their less conspicuous counterparts in Texas or Alabama.

Army ants could even serve a more prosaic function, and one potentially useful to humans. Adrian Smith and Kevin Haight from Arizona State University pointed out that because other ant colonies tend to flee with their brood, evacuating the entire colony, queen and all, when army ants approach, researchers attempting to collect the prey species could exploit this behavior and save themselves time and trouble. They cheerfully—and in my opinion, just a tad cold-bloodedly—suggest that fellow scientists use a small batch of army ants to rout a desired species from their subterranean chambers, rather like tiny terriers sent after rabbits in a burrow, saving hours of often back-breaking labor excavating the nest. Their paper even includes links to videos that demonstrate, in eerily infomercial fashion, exactly how nifty and efficient this technique can be. It sounds like a great idea. But somehow "terrier ants" doesn't have much zing.

Adopting the Enemy

If the army ants are more hunters stocking their family larder than noble soldiers proving their mettle, what about the slave-making ants with their "wonderful" instinct, as noted by Darwin? The wars, or raids, that these ants undertake are not about getting food, at least not directly. A slave-maker colony is started when an inseminated queen of one species enters the nest of another, kills or expels the resident queen or queens, and begins to lay eggs of her own kind. Her children are then reared by the host species, which accepts them as if they were nest mates. To replenish the host workers, the slave-maker species sets out on periodic expeditions to snatch the larvae and pupae from another host colony, bearing them back to the slave-maker nest to mature and act as normal workers for their hosts. About fifty of the eleven thousand known ant species behave in this manner, with some capable of living on their own and others so specialized that they cannot even feed themselves without the help of their captives.

Despite their rarity, the slave-making ants have attracted a great deal of human attention. Charlotte Sleigh documents the fascination of nineteenth-century observers with the slave-making ants, many of whom were quite overt in their analogy with the human slave trade of the time. Perhaps surprisingly, many naturalists and authors decried, not the practice of exploiting the labor of others, but the "degeneracy" of the slave makers themselves. In characteristically opinionated prose, Maeterlinck disapprovingly notes that the slave makers "cannot eat without assistance, for they cannot take any nourishment save from the mouths of their servants. They are as little capable of rearing their young as of building or repairing their nest. In the depths of their lair they pass their time in besotted idleness, rousing themselves only in order to polish their armor, or to pester their slaves for a mouthful of honey. Without their servants these magnificent warriors, with their bronze armor, these superb shock-battalions, these irresistible veterans of great campaigns, are as impotent, as utterly helpless as so many suckling infants." In his 1954 book *Ways of the Ant,* John Crompton was similarly censorious: "Even if their slaves do not desert them, mental and physical decay will in itself and in its own time exterminate them. There must be many species of slave-making ants that have died out for this reason."

Although his language is rather histrionic, Maeterlinck was scientifically accurate, at least with regard to the obligate slave-making ants; in the early 1800s the great entomologist Pierre Huber had placed a group of ants of one of the slave-maker species in a kind of ant farm, along with honey to eat and some of their own pupae and larvae. Within a few days half had already died and the remainder were on the brink of starvation.

The raids themselves can be quite dramatic to witness. The species whose brood is being taken generally attempt to drag the pale, helpless larvae and pupae away from the nest, only to be pursued by the workers of the host ant species. Raids seem to be confined to certain times of the year, and at least some of the ants studied in this regard use cues from within the nest to decide when to begin raiding behavior. Slave making in ants is confined to temperate regions of the world, and scientists have suggested that the absence of seasons in the tropics explains the lack of raiding and, hence, slave-making behavior, since there is no internal signal that indicates when pupae can reliably be abducted from their nests. Some species also raid at certain times of day. Joan Herbers, a scientist at Ohio State University and an authority on such ants, says that when she was at Colorado State University her students knew exactly when to go looking for raids: "Jeremy [her student] could head to the hills of Colorado around 10 in the morning, knowing he would be done with fieldwork by 3 or 4." Others are not so reliable: "We have set up many experiments in the lab; some days they raid and other days they don't. Some days they raid fiercely and other days the raids fade away. Sometimes it takes an hour and other times 6–8. It's a pain, and has frustrated several journalists who have visited my lab."

In addition to studying the ants as a scientist, Herbers has questioned the wisdom and accuracy of the name *slave-maker ants*. She is not alone in this regard; Hölldobler and Wilson point out, "It is traditional to use the expression *slavery* for the exploitation of one species by another. In the human sense this is not slavery but more akin to the forcible domestication of dogs and cattle by humans." They go on to detail situations in which ants use the labor of others from the same species, but the term *slavery* is clearly limited in its applicability. Some entomologists use the more technical jargon term *dulosis* to refer to the process, whether within or across species, but most scientific journals still call it *slave making*.

Herbers is not just concerned about the use of the word *slavery* by scientists. She questions its suitability given its obvious connotations of human activity. At public lectures, she often is asked about the parallel between ant and human slavery, a parallel she always decries. She has come to the conclusion that we would all be better off abandoning the metaphor and terminology entirely, because of its emotionally loaded overtones. As an alternative, Herbers proposes the term *pirate ant*, since human pirates also make raids and steal cargo, often killing some of the victim ship's crew. Scientists could continue to discuss raiding parties, captives, and booty, without recourse to the loaded terms that certainly bring the public up short. I am in agreement with the distaste for the word *slavery* in nonhumans, and use it here only when the original authors use the term, so as not to rewrite their usage.

Regardless of its social baggage, however, another problem with calling the ants slave makers is that, as with the army ants, it gives an entirely incorrect view of what the ants themselves are doing. Hölldobler and Wilson's point about domesticated animals versus forced labor from members of the same species aside, most biologists, including them, classify the behavior as a kind of parasitism. In other words, the so-called slave makers are acting like exceptionally free-roaming tapeworms. Like the tapeworm, the slave makers, at least the obligate species, make their living entirely off of another organism, the host. But instead of traveling passively from one host's intestinal tract to another via, say, a contaminated bite of meat, the ants take matters into their own six legs. The slave raids, with the excited workers rushing to and fro with their cargo of pale cocoons, are just a more visible and dramatic version of the worm in the gut ensuring it will have someone to provide it with a steady supply of meals for the foreseeable future. Even Crompton notes that "a slave-raiding expedition is not really a battle, it is a routine commercial undertaking."

Admittedly, this analogy is not perfect, and the ants are what scientists call *social parasites,* rather than internal ones. Cuckoos and cowbirds are the most familiar examples of such animals: the cuckoo female lays her eggs in the nest of another bird species, exploiting the parental behavior of the host, who rears a genetically unrelated chick. The hosts have, in a sense, adopted the enemy to their own advantage, gaining the labor of others at little expense. And Maeterlinck points out that the captive ants do exactly the same thing that they would be doing in their own nest, namely, feeding the workers and caring for the queen. Their lives are no harsher than they would be in their own nest, and the everyday life of any ant is pretty grim by anthropomorphic human standards at least. But those pejorative declarations about degeneracy from Crompton and Maeterlinck fit right in with this point of view. Tapeworms and many other parasitic organisms have reduced limbs, eyes, and other organs, a state of affairs that probably evolved because the appendages are unnecessary, maybe even an impediment, in the dark cozy confines of the host's gut. Crompton's prediction about the extinction of the slave makers may be off the mark, since of course parasites show no signs of going out of business. Seeing slave making as a form of parasitism gives rise to the unsettling thought that, by the same token, we are a kind of slave to our own pathogens.

Viewing the interaction as parasitic not only sidesteps the terminology melodrama, it clears the way for asking other interesting questions. Herbers and her colleagues have examined variation across the range of several species of pirate ants regarding which species they exploit, and, as with a disease-causing organism, talk about the "virulence" of different raiding species. Just as anthrax is more virulent than athlete's foot, by doing more damage to its host, a more virulent social ant parasite kills a larger proportion of the adult ants at the nest it raids.

With postdoctoral scholar Christine Johnson, Herbers introduced two different slave-making species that parasitize the same host species into outdoor enclosures in a field in Ohio. The enclosures had one or the other slave-making species or both at the same time, along with the host species. The researchers then waited to see how the host species did, predicting that the presence of both slave-maker species would be the biggest burden on the host ants. Much to their surprise, the host colonies did better when both parasite species were present together. Johnson and Herbers speculated that the two types of slave makers might have competed with each other, to the detriment of both, leaving the host ants to prosper unmolested. This kind of complicated interaction among several species is becoming increasingly interesting to scientists, since it suggests that we need to look at more than just one species at a time to understand an animal's ecology. The researchers concluded that variation in the abundance of slave makers could affect "hot and cold spots" of ant abundance in the forests where the ants occur.

Just such geographic variability in ants was the subject of a study by Susanne Foitzik, now at Regensburg University in Germany but formerly another postdoctoral scholar working in Herbers's laboratory. Foitzik and others have recognized that the ants are a good way to study ways that a host and parasite can influence the evolution of each other, in what's called a coevolutionary arms race. After all, one wouldn't expect the host, or exploited species, to just sit back and take it — for example, we evolved an entire immune system to resist the attacks of viruses and bacteria. Other kinds of hosts of social parasites show varying degrees of defenses against the parasite; some cuckoo and cowbird hosts recognize and reject the interloper's eggs, while others seem to be oblivious to the gigantic size of the parasite chick relative to their own offspring and valiantly stuff food into the cuckoo chick's gaping maw at the expense of their own reproduction.

Foitzik and her coworkers looked at the ways that the slave, or host, species varied in its ability to defend itself against the slave makers. They were interested in whether the defense mechanisms were the same in different places, regardless of the intensity of the raids by the slave makers, or whether each pair of host and parasite populations evolves a unique way of interacting, with a new arms race in each locale. They compared colonies of a raiding species and its victims in the Huyck Preserve in New York state with those in West Virginia. More and larger colonies of the slave-making species occur in New York, which should make the pressure on the host species more severe, since they are being raided more frequently. The slave-making ants in turn can kill the queens of their hosts without too many repercussions, since many colonies of potential victims are also present.

The scientists found that the coevolution between host and parasite was in fact different in the different places; in New York, a guard ant was more likely to be found protecting the host nest entrance, and in turn the New York slave-making ants took more of the brood from the nests they raided. The host defenses were also more aggressive to the initial scouts sent out by the raiding parties. "Ironically," write the researchers, "these host ants are probably killed by enslaved conspecifics [members of the same species] that accompany . . . workers on raids, rather than by the slave-makers themselves." The defenses, however, weren't unique to a particular set of nests, supporting the idea that universal defense mechanisms evolve throughout the population.

The idea that the hosts could defend themselves against the raiders wasn't given much credence until recently, and it's tempting to speculate that the lack of exploration of the idea came from people clinging a little too tightly to that slavery analogy. Slave rebellions are risky and scarce. But it's commonplace to imagine a host and parasite, for example, the worm inside the gut of a mouse, continually evolving ways to attack or defend against each other.

Whether you think of it as piracy, parasitism, or slavery, capturing live individuals of another species and benefiting from their labor requires a complicated set of behaviors. How did such a practice evolve? Charles Darwin offered the first potential explanation in *The Origin of Species*, proposing that the ancestral slave makers first took the pupae as prey. When some of the pupae accidentally escaped detection back in the host nest and became adult workers, they were not perceived to be foreigners and, hence, began doing their normal ant activities, which made the colony as a whole prosper.

Another possible route to the evolution of piracy is via the territorial battles that commonly take place between colonies of the same species. Ants and other social insects usually have very strong loyalties to their own colony and will attack intruders that smell like they come from a foreign nest. If a new colony is established too near an existing colony, fights between the workers of the two groups can result, and several scientists have suggested that this generally pugnacious behavior could have evolved to be directed at ants of other species as well. If the two species were closely related, and hence shared a more recent common ancestor, the likelihood of them also becoming tolerant of a captured pupa or larva is increased, because the captive would smell more familiar.

Jeannette Beibl, a researcher at Regensburg along with Foitzik, examined the DNA of numerous slave-making species. They and colleagues R. J. Stuart and J. Heinze determined that the practice evolved independently several times in different groups of ants, some relatively recently, at least by evolutionary standards. This variation suggests that different selection pressures might have caused slave making to evolve in the different species.

Six-Legged Constables

IF THE army ants aren't a real fighting force and the slave makers are just parasites with an uncanny resemblance to their hosts, do treachery and aggression exist at all for the social insects? The answer is a resounding yes. The carnage is subtle, but far more devastating in its after effects than even the most formidable slave-taking raid. In evolutionary terms, loss of life is not nearly as injurious as loss of reproduction. The social insects, with their suicidal colony defense and sterile workers, have perplexed evolutionary biologists since Darwin. While biologists have mostly explained the benefits of such extreme cooperation for the individual colony members, those busy little virgin bees and ants still turn out to show some enterprising forms of rebellion.

Although worker ants, bees, and wasps cannot mate, they often possess functional ovaries and can produce their own eggs. These unfertilized eggs develop into males, because throughout this group of insects and a few others, daughters have two copies of each chromosome, like humans and other vertebrates, and develop from fertilized eggs, but sons have only the mother's genes and, in effect, have no father. (I often give my animal behavior students an exam question that asks whether it is true that a honeybee male has a grandfather but no father. The ones who get it are triumphant, often hammering the point home in far more text than necessary, while the ones who don't flounder in ever-widening circles of confusion; one ended up declaring in apparent despair, "No, every animal has a father, if they didn't have a father they wouldn't have a mother and then what would happen?")

This genetic oddity means that workers are often more closely related to each other than they are to their own offspring, although the exact proportion of shared genes varies depending on how many males have mated with the queen. The real genetic payoff comes, as I pointed out in an earlier chapter, not from the workers helping to rear their sterile sisters, but from production of the future reproductives, the queens and drones that will leave the colony and found a nest of their own.

Under some circumstances, therefore, it is beneficial to an individual worker to lay some eggs that will become male reproductives. But the other workers would pass on more of their genes by investing in their brothers, the sons of the queen, rather than their nephews, the sons of their sisters. So you might expect that workers would sabotage each others' efforts to slip a few of their own eggs into the hive. Indeed, Francis Ratnieks and Kirk Visscher documented exactly such behavior, termed *worker policing,* in honeybees. The bees are able to tell which eggs are laid by the queen and which by their sister workers and will remove the latter and prevent them from developing. Visscher and Reuven Dukas discovered that the workers can even detect the degree of ovarian development in their sisters and act more aggressively to the ones that are on the verge of producing their own eggs.

Ratnieks, along with Tom Wenseleers, took the idea of worker policing further. They pointed out that the better the workers are at stopping each other's attempts at reproduction, the more likely it is for workers to give up, in effect, and simply put all their efforts into the queen's offspring rather than try to produce their own. To test this idea, the scientists compared the proportion of egg-laying workers in ten species of wasps and the honeybee; the insects vary in the effectiveness of worker policing in the nest. As they predicted, workers from species in which the policing is stringent are much less likely to try to lay their own eggs in the first place. The scientists conclude that the insects "provide evidence for something that has proved notoriously hard to demonstrate in human society: that better law enforcement can lead to fewer individuals behaving antisocially."

Ratnieks and Wenseleers also noted that the workers can control each others' reproduction in a different way, by regulating which female eggs end up as queens and which as workers. In many, though not all, social insects, this caste difference is determined during development, with future honeybee queens, for example, placed in larger cells than the plebian workers and fed more of a special substance called royal jelly that jump-starts their growth. Reproducing oneself, rather than caring for the young of others, is an attractive evolutionary prospect, but developing into a queen is only part of the process. It's rather like becoming a movie star: being stunningly beautiful, while essential, is no guarantee of red carpet status. Only a tiny fraction of the queens produced will actually make it to the Oscar-winning equivalent of starting their own hive. But it's not good for the colony if too many individuals become queens, because queens don't do any of the foraging, cleaning, or other mundane tasks of daily life. And yet, as with eager celebrity wannabes, the starlets rush to audition. As Ratnieks and Wenseleers put it, "The lottery to reproduce is so attractive that many more enter than could possibly win the prize of heading a new colony." Policing by the other workers prevents too many queens from being reared, because the larger cells in the comb for rearing queens are strictly limited.

A tropical stingless bee called *Melipona* provides an elegant illustration of the scope and limitation of policing. Unlike honeybees, which are reared in wax cells that are open at the top so that the workers can feed the larvae a bit at a time over their development, the stingless bee queens are about the same size as workers and are reared in sealed cells, each of which contains its own ration of food. The female *Melipona* thus develops into an adult without interference from other bees and can become either a queen or a worker. As a result, up to 20 percent of females are aspiring queens. But grim reality sets in once they emerge from their virginal chambers: lack of policing beforehand means that many of the new queens are set upon by the workers and torn limb from limb. The policing ameliorates this carnage by preventing too many queens from being produced in the first place.

Punishment of cheaters who try to reproduce on their own in a social insect colony is not confined to bees. Ordinarily, only queen ants produce a particular chemical on their body's surface to indicate their reproductive status. But if a worker's ovaries develop and she begins to lay eggs, the other workers detect the same odor on her body and attack their sister. Adrian Smith, Bert Hölldobler, and Jurgen Liebig painted workers with the telltale compound and induced the aggression, showing that the odor is indeed the trigger for detection of cheaters. In a colony with its queen removed, however, the newly reproductive workers are left alone.

Ratnieks and Wenseleers ask, "Can humans learn anything from insect policing? The principal lesson seems to be that policing is a common feature of social life and helps to resolve the conflicts caused by the transition from individuals to societies. . . . Policing in human societies has been used by repressive regimes to sustain inequalities, as demonstrated by the negative connotation of the phrase 'Police State.' But a human society in which policing is used to promote greater equality and justice may not be an unattractive prospect." Of course, the conflict between the good of society and individual freedom is an old one, and not likely to be settled by observations from the beehive. My take on the sinister world of Big Sister is that such behavior is far more deadly than the army ants swarming over every living thing in their path. Who needs nuclear weapons?

CHAPTER 9

Six-Legged Language

The Battle between Impatience and Procrastination

SUDDENLY, a lot of bees seemed to be flying in and out of our garage. There has always been a space of about a foot between the top of the wall and the roof, so it is not an enclosed building, but while I was used to seeing insects, our cats, and the occasional possum or raccoon making use of the opening, for a couple of days a steady stream of humming golden-bodied workers went back and forth at one corner. Needless to say, no flowers grow in our garage, so I wasn't sure why I was seeing half a dozen or so bees buzzing overhead every time I went to get my bike.

I told my husband. They're just bees, he said. Don't worry about it. I wasn't worried, exactly. I just was a little leery of what the increased traffic might portend. And on the third day I noticed a baseball-sized clump of bees surrounding a beam inside the garage near the opening they had entered. Uh-oh, I thought. They're swarming.

After a suitably pointed "I told you so" to my husband, I knew just what to do. I called Kirk Visscher, who is a professor of entomology at my university and an expert on bees and beekeeping. "Help," I said. "We've got bees." Kirk drove over to our house with a portable hive constructed from boards, which he set out at what we hoped was a tempting yet suitably lengthy distance from the beam, and our garage. With a little luck, he assured us, the bees would move their operations into the hive and he could take the colony to use in his research back on campus.

Luck, however, was not with us. A day later, the clump had grown to the size of a small football, and honeycomb was visible when the layer of bees shifted. Kirk returned, this time with the age-old beekeeper's equipment of a funnel and smoker. He gently pumped the smoke over the cluster, which had no discernable effect on the bees other than to make them hum in what, at least to my ears, was a rather agitated way. I backed up into the driveway and asked Kirk exactly what we were trying to accomplish, since having a garage full of annoyed and active bees did not seem like an improvement over having one full of reasonably quiescent ones.

"We're just trying to convince them that this wasn't the great place they first thought it was," he said. Like moving to a neighborhood you later discover has bad air quality or a lousy school system, I guess, only without the complication of already having filed the paperwork and commencing escrow. Whatever the rationale, in another day the entire swarm had suddenly decamped, as if they had never been, except for the telltale irregularly shaped sheets of fragrant honeycomb attached to the beam. Presumably, the bees had left for a more permanent home, whether in a tree cavity or some other fortunate beekeeper's hive, but at any rate sufficiently far from our garage.

Honeybee swarms are the colony's way of reducing overpopulation. When a colony grows large, the worker bees nurture new queens, and the old queen, along with perhaps half of the workers, goes off to establish a new hive somewhere else. Initially, the pioneering bees and their queen settle in a mass like the one inside our garage, while scout worker bees go out and look for a new place to live. For centuries, enterprising humans have taken advantage of these temporary swarms to capture a new colony and place it into a manufactured hive, as Kirk had attempted. While handy for the beekeepers, this abrupt human-induced end to the house-hunting process had meant that no one paid attention to, or understood, the extraordinary process the bees undertake in making a decision about where to go.

Remember, the bees are in a swarm of up to ten thousand individuals, surrounding a plump, fecund queen, who must be protected at all costs. Only a few bees are able to survey their surroundings and choose a site where they and subsequent generations will spend the rest of their lives. What makes for exclusive bee real estate? How do the scouts convey the news about where they have been to everyone else in the swarm? And once the information is delivered, how does the group decide which of the prospective new homes is the most suitable? Finally, how can such an enormous group of tiny creatures stay together and get to the same place?

The answers to these questions shed light not just on the behavior of bees, but on how decisions are made by groups of animals, whether these are insects, migrating birds, or humans. At the same time, only bees and their close relatives show the kind of complex communication system that challenges our definition of what it means to be human. How do insects decide where to go? Further, and more provocative, is the means by which they indicate their destination a real language?

The Spirit of the Hive

ALL KINDS of animals, and even some microorganisms, make decisions: go left instead of right, eat this food and not that, sing now or rest for the afternoon. Decisions by female *Drosophila* about where to lay their eggs have vital implications for the fate of the offspring, and scientists have made great strides in understanding how genes control the fruit flies' choices from among an array of options (too much sugar in the medium and they turn up their little ovipositors, or egg-laying organs, at it). But the flies do not have to consult with their families about their decisions, and no one else offers an opinion.

The bees, however, live in a society, and while it is not democratic, neither is it a dictatorship. The queen may have the last word in reproduction, but not in moving house. Maurice Maeterlinck, the playwright who praised the nobility of ants and their cooperative nature in his 1930 book, had written *The Life of the Bee* in 1901. He considered the way the colony found its way to its new home, and concluded, "All things go to prove that it is not the queen, but the spirit of the hive, that decides on the swarm." As with house hunting in humans, the decision about where to move in honeybees and other social insects is fraught with complication. The decision must be made fairly quickly, because the swarm is vulnerable as it clings to a branch (or garage ceiling beam). At the same time, its consequences are crucial, since the colony will spend its life in its new home and needs to have ample room to raise its brood, with food sources located nearby. How do the bees keep from spending endless time in filibuster and argumentation, like a miniature all-night congressional attempt to arrive at a budget? What is more, sometimes the old home is destroyed by fire, flood, or the untimely arrival of a hungry bear, necessitating the abrupt evacuation of the old home and a pressing need to find a new one. A group decision is essential; the bees can't simply go back to their constituents and try again next season.

Group decisions are particularly interesting because they imply first that the members of the group are able to convey messages to each other, and second that they have some mechanism for evaluating each individual's contribution. Groups have a rather dubious reputation when it comes to collective activity; none other than Friedrich Nietzsche disparaged humans by suggesting, "Madness is the exception in individuals but the rule in groups." Maeterlinck, however, was more charmed by the process of group decisions in the bees and felt that "there can be no doubting that they understand each other," although "certain as it may seem that the bees communicate with each other, we know not whether this be done in human fashion." The bees exhibit what is called a consensus decision, which means that they choose between several mutually exclusive alternatives and then all abide by the selected one. This process is similar to what goes on in a democratic election, or in international treaties signed by a number of nations with a common goal. Everyone doesn't necessarily contribute to the decision, but they all agree to do the same thing in the end.

Consensus decisions are distinguished from what are called combined decisions because they require everyone to concur, which means the group has to possess a fairly sophisticated means for exchanging information. A combined decision is made when the individuals in a group simply assign themselves different tasks, for example, the allocation of hive cleaning versus foraging in bees, but don't all agree beforehand on a list of who will do what. The distinction is important because getting everyone to agree to a single outcome means that they may have to sacrifice their own interests in doing so, a rather advanced capacity for a tiny insect. Many scientists first started thinking about the possibility of a "hive mind," that controversial greater-than-the-sum-of-its-parts superorganism, when pondering the way that social insects choose a new place to live.

Aristotle had noticed that it seemed as if bees received information from scouts that had made advance sallies to find potential nest sites, but the process of deciding on a new nest site was first studied in detail in the 1950s by a German zoologist, Martin Lindauer. He happened upon a swarm near the Zoological Institute at his university in Munich and noticed that a few of the bees on the outside of the swarm were performing the same waggle dance that his mentor, the Nobel laureate Karl von Frisch, had described in the context of signaling the whereabouts of food sources. Since the bees had no pollen or nectar, Lindauer wondered if they might be signaling not the location of a patch of flowers, but the possible place the swarm might settle.

By recording the different locations encoded in the dances of the apparent scouts, Lindauer noticed that although the bees seemed to be dancing "for" many different sites at first, eventually they seemed to settle on just one. Soon after this winnowing of alternatives, the swarm rose in a body and took off for the site that had made it to the finals. In the decades that followed Lindauer's work, scientists established the characteristics that bees would use in their description of a dream house to a bee real estate agent, including a south-facing entrance, a small enough entrance hole to discourage unwanted visitors, and enough room for an average-sized honeybee colony to spread out in comfort. The ability to compare several different possibilities, like enterprising couples scanning the online real estate ads, indicates a rather sophisticated cognitive ability on the part of the insects and has even led to the suggestion that the bees possess some form of consciousness. Why being able to choose a split-level ranch over a refurbished Victorian is peculiarly emblematic of higher intelligence, while other decisions are not, is unclear to me, but it is undoubtedly a complicated decision.

In the 1990s, Kirk and his former advisor, the eminent bee expert Tom Seeley at Cornell University, began to work with other colleagues to determine how the selection of just one site was made. One swarm took about 16 hours of dancing, spread over three days, to reach a decision, with eleven different potential sites taken under consideration before the winner was determined. In a *New York Times* column, James Gorman noted that the scientists were, like Maeterlinck, convinced that the bees arrived at decisions that were good for the group, and that "Dr. Seeley is a bit more cheery than Nietzsche," a comparison that was probably novel for both the entomologist and the philosopher.

So how do the bees choose the winner? The idea of a bee version of polling the hive constituents to get a sense of everyone's views on the various sites before arriving at an informed decision is appealing, but it turns out that this is not how the bees decide which scout to follow. Instead, they seem to be sensing a quorum of dancers "for" a particular site, and then the entire swarm follows the quorum.

Kirk, Seeley, and Kevin Passino of Ohio State University made this discovery using bee colonies on Appledore Island off the coast of Maine, where Cornell has a research facility. The island was handy because it has almost no trees that could serve as natural nests, which means the scientists could provide all the potential homes for the colonies they brought there.

The scientists gave different swarms of bees on opposite sides of the island either just one nest box into which they could relocate, or five similar boxes set close together. If the bees required a quorum of dancers, the dilemma of five equivalent alternatives should delay the formation of the quorum and, hence, delay the swarm movement itself, but the rest of the decision-making process would be the same in the two cases. As predicted, the bees took an average of 442 minutes to arrive at a decision when they were spoiled for choice, as it were, compared with just 196 minutes when one box was available. "Group intelligence," the researchers concluded, "is a product of disagreement and contest." Whether that constitutes more optimism than Nietzsche is perhaps a matter of debate.

Kirk and Tom Seeley also determined that the bees dance differently depending on the quality of the site they have discovered; scouts spent equal amounts of time inspecting two potential sites that were offered, but performed more circuits of their dance back at the swarm for the better location. The bees also avoid getting drawn into agonizing fruitlessly over a poor candidate by the rapid decrease in sequential visits to a lower-quality site, so that they are able to ruthlessly reject a loser instead of second-guessing themselves, something more humans would probably do well to emulate.

How the bees assess the presence of enough scouts to constitute a quorum is still not well understood. The swarm does not always arrive at a unanimous decision; occasionally one will split at takeoff, and even when dissention is not so drastic, a few divergent dancers will still be rooting for their own selection up until the very last minute. Until very recently, scientists likewise could not understand how the group of bees could all get up and move in synchrony after the winning site was selected. The scouts produce a kind of rallying cry, called *piping*, that seems to energize the swarm to warm up before they take off in unison.

Some work by Seeley and Clare Rittschof of the University of

Florida suggests that the scout bees work the crowd by moving among the more languid members of the group, making a stereotyped set of motions accompanied by sounds, called a *buzz-run.* The buzz-run seems to encourage any laggards to move their wings as well, which in turn ensures that everyone's wing muscles are sufficiently warmed up for flight. Because they are cold-blooded like other insects, bees need to reach a certain temperature before they can fly, and they do so by revving their muscles like diminutive engines. Rittschof and Seeley then proposed that the scouts act like swarm thermometers to gauge the temperature of the mass of bees and trigger its synchronized takeoff when everyone is ready.

Even after takeoff, the synchronicity of the swarm is amazing. Although the scouts have done their advertisement and the decision has been made, less than 5 percent of the bees in the swarm have actually been to the site themselves and, hence, don't know exactly where to go. Yet all of the ten thousand or more end up in the right place, often miles away. Another publication with Seeley as the bee expert, this time with engineer Kevin Schultz from the Ohio State University as a collaborator, gave the solution to this problem as well. Using high-definition films of the swarms on Appledore Island, the researchers examined more than 3,500 frames of bee movement and determined that the informed few fly through the top half of the swarm at high speed to lead the way. These leaders are called *streaker bees,* after their cometlike movements in the cloud of insects.

Microscopic Real Estate

PEOPLE have long taken a personal interest in the real estate preferences of honeybees, mainly because the product of a smoothly running hive is relevant to human well-being. They have been less concerned about the decisions made by other social insects, such as ants, so long as they do not decide to relocate in or near human habitation. The house-hunting behavior of ants, however, is both similar to and different from nest site selection in honeybees, and the contrast is instructive.

Much of the research on group decisions by ants on where to live has centered on a couple of species with life histories that sound like they come from a fairy tale, or maybe a Winnie the Pooh story. The insects are even tinier than the ants commonly seen in kitchens, and use either rock crevices, or, more charmingly, acorns, as their place of residence. The entire colony, queen and all, can fit into a space smaller than a person's thumb. Needless to say, this makes replicating their world in the laboratory extremely easy; one of the foremost researchers on these ants, Nigel Franks of the University of Bristol, makes little homes for them by gluing a bit of cardboard between two glass microscope slides to recreate a crevice that is also easily spied upon.

The ants' house-hunting activities have attracted attention mainly because they use the tandem running procedure I described in the chapter on learning to help their nest mates find the new location. As in the bees, ants send out scouts to search for new homes, but unlike the bees, the scouts enlist enthusiasts for the new cavity by the same "follow me" motions used to direct other colony members to food sources. The ants don't perform waggle dances like the bees, but they will recruit for good sites more quickly than for poorer ones. The new recruits, if they concur with the desirability of the chosen location, then get others to join them. If you are an ant living in a small dark hollow, the best home has a narrow entrance with dim light, presumably to discourage predators, and just the right amount of floor space. The ants can even evaluate the potential for nasty neighbors, in the form of a foreign ant colony, and eschew such potentially troublesome locations.

At this point the process diverges from that of the bees, because once a site is selected by enough ants, the process becomes one of shanghai rather than persuasion, as the remainder of the colony is simply picked up bodily and carried, head ignominiously pointing backward, to the chosen location. Carrying is three times faster than tandem running, and once the carrying starts, the ants are committed to their new location and do not switch preferences in midstream. Regret, it would seem, is not part of the ant repertoire. As a paper by Robert Planqué and his collaborators, including Franks, puts it, "Ant colonies have found a good compromise between impatience and procrastination." Would that we were all so prudent, at least when it comes to moving house.

In human groups it's often the case that the larger the crowd, the harder it is to reach a decision, whether about going to war or choosing a restaurant. In contrast, the ants seem equally good at determining the best nest site from among an array of options presented in the laboratory regardless of the size of their colony. The larger colonies do use more of the tandem runs to exhort others to follow them, and seem to need a larger quorum of assenting individuals before they decamp. Interestingly, both small and large colonies select sites that will comfortably contain a colony that has grown to full size, suggesting that the ants can anticipate their future needs, a remarkable feat.

Franks and his colleagues demonstrated this ability of ants to plan by showing that they can distinguish between various qualities of nests and perform reconnaissance when they evaluate the possibilities. The official definition of reconnaissance, according to the *Oxford English Dictionary*, is "an examination of a region to ascertain strategic features through a preliminary survey," and the ants manage this by retaining information about the different potential nests and using it later to recruit for different sites. Like a person who sees an ice cream store while jogging in the morning and files away the information for dessert, the ants can remember the landmarks near a potential nest site or the odors left there, even if they are not actively house hunting at the time. If these cues are removed by an investigator, and the ants are then required to find a new home by virtue of the scientist destroying the old one, the ants can no longer distinguish among the sites.

Ants can also arrive at the best group decision almost as a byproduct of individual behavior, without the need for extensive communication. Say that a tasty bit of food can be found at the other side of a deep crack in the ground from the colony. A leafy branch lies across the crevice, and the ants can either take a shorter, direct path to the food or a longer, more convoluted one, depending on which twig they use to cross. The shorter path would be more efficient, and it turns out that this is the one favored by the ants. But how did they arrive at the decision? Even for a devoted myrmecophile, it defies reason to imagine the ants testing out one path and then the other, timing both, and then sending the message to the rest of the colony that they can save their exoskeleton some wear by taking the shorter journey.

It turns out they don't have to. Franks and several other scientists determined that a much simpler process is at work. As an ant returns from a food source, she lays down an odor trail that attracts her nest mates. The more ants that have been back and forth from the colony, the stronger the attraction of a particular pathway. Thus, the shorter trip over the twig gets more use and builds up more odor, simply because it takes less time to go to and from the food, and the ants themselves reinforce the easier path as the best choice. Others follow and, voilà, the colony as a whole has made the right decision. Similar behavior allows the ants to select the easiest sites to excavate when the possible nest entrances are blocked with sand.

At least one other species of ant, the delightfully named gypsy ant, can make collective decisions about which kinds of food can be harvested singly and which require enlisting reinforcements. A group of French and Spanish researchers presented the gypsy ants with dead crickets, which could be moved by a cooperating group of ants but not by a single individual; dead shrimps, which are five hundred times heavier than a worker and must be butchered into individually transportable pieces; or sesame seeds, which as any picnicker knows can be easily borne aloft by an ant acting alone. Making off with food in a timely and efficient manner is important, because other ant species are potentially lurking nearby, ready to snatch any food left unattended. The ants were able to gauge the number of workers necessary to lift and carry the crickets depending on the size of the prey, with small crickets requiring about a dozen workers but large ones fifteen or more, and they quickly recruited an even greater number of ants to carry out the dismemberment of the shrimp before it could be detected by competitors.

Not all group decisions by insects have such a happy outcome. Although the social ants and bees get most of the attention, scientists have also examined collective behavior in forest tent caterpillars, which live in groups until they spin their cocoons and become adult moths. The caterpillars move in munching hordes through the treetops and may either linger on a particularly succulent tree or move quickly through it in search of a more nutritious set of leaves. Because of the caterpillars' discerning tastes, forests that have been attacked by the caterpillars are often a patchwork of ravaged and intact trees. In nature, experiments have shown that they prefer carbohydrate-enriched or untreated aspen leaves, rather than leaves with a high protein content, a kind of anti–Atkins diet.

Offered a choice between diets concocted in the laboratory that differ in their nutrient composition, an individual caterpillar will make the "right" selection and eschew an unbalanced low-carbohydrate food in favor of one with the natural blend of proteins, carbohydrates, and fiber.

In groups, however, the caterpillars, like schoolchildren egging each other on to eat Doritos and Twinkies instead of carrot sticks, will often end up choosing the less nourishing offering. The problem seems to occur because, like ants, the caterpillars follow odor trails left by their companions. The initial decision to taste one or another of the foods is made at random, but once a caterpillar has started eating, its odor trail encourages others to follow, and then the entire gang gets trapped by heeding the message that went before it. The caterpillars thus follow each other to their collective nutritional doom. Unlike the bees or ants, the caterpillars lack any capability of communicating their state to each other, so they cannot indicate that they have arrived at a less tasty branch and warn others of their folly. Nietzsche's pessimism about groups seems to be better illustrated by the caterpillars than the bees in this regard, which makes you wonder whether we are so close to the social insects after all. Luckily, the caterpillars differ in their tendency to move around, and if a large proportion of the group was of a more active predilection, the group itself was less cohesive and managed to escape the poor decision.

Flying versus Walking, and the Lead-up to Language

DECISIONS about food or nest sites are closely tied to the success or failure of any given insect colony, and the way that different species get help when an individual finds a food item that is too big for it, or needs to get everyone else on board with a decision has important implications for social behavior. The way that an insect recruits is, in turn, constrained by its own biology. Ants, as I've discussed, lay down an odor trail that becomes stronger and stronger as more workers use it, but to each ant that traverses the trail, its end point is a mystery — she simply follows her nose, so to speak, until she reaches the goal. In the case of establishing a new nest, ants can be carried by their nest mates to the new site, and again those being unceremoniously tucked under a leg need have no idea of where they are being taken.

Bees are different. Flying instead of walking means that you can't easily haul your sister workers around, which means that the bees need some other way to convey information to the rest of the colony. And although some species of bees do place dabs of odor on plants and other objects as signposts on the way to a food source, pheromones are not nearly as satisfactory a method for indicating directions for flying insects as they are for crawling ones; the bees have to continually dart down to the vegetation, and the odors fade without continual reinforcement by a stream of workers.

What's more, at least some kinds of bees have to worry about eavesdroppers on their odor cues. James Nieh at the University of California at San Diego has been studying tropical stingless bees in Brazil, Panama, and other parts of Latin America for many years. The stingless bees are social, like honeybees, and Nieh noticed that the species he was studying left scent marks near good food sources. The problem was that the scent marks were easily detected by a larger and more aggressive species of stingless bee, and when the bullies found the food, they dispatched their victims with what Nieh describes as "a range of forms of aggression from threats to intense grappling followed by decapitation." The victim species avoids the odor marks left by the aggressor species and sticks to its own signals, but the aggressor does the opposite, preferring the odor marks of the victim species to its own.

What's a bee to do? One possibility is to encrypt your directions. Instead of setting out an odor that broadcasts "Tasty morsel here!," whisper your findings only to those for whom the message is intended: your nest mates. In other words, evolve a symbolic language with which you can convey what you know to others in the privacy of the hive, without fear of being overheard. Nieh suggests that the famous dance language of the honeybee, and its counterparts in a few other species, evolved under pressure to hide indications of the location of rich food sources from any competitor bees in the area. Ideally, of course, one would have a code able to be read only by the members of one's own colony, but that degree of encryption seems to be beyond the bees, and so they have had to settle with having species-specific, or at least population-specific, signals. Combined with the other advantages of such communication to a flying rather than walking insect, for example, the inability to carry other workers and the inconvenience of odor trails over long distances, the dance doesn't seem like an anomaly, but like an obvious solution to a problem.

Ants and bees differ in a few other respects: ants are much slower than bees at redirecting their efforts to a newly introduced rich food source, and the members of an ant colony act almost like neurons in the brain when responding to stimulation. Ants also exhibit something called *stigmergy*, which sounds like either an eighties band or what happens when the recipients of social stigma gather in groups, for example, smokers outside a building, but is the way that the ants coordinate each other's movements by changing the odor trails that convey activity patterns. This too means that the ants can make decisions without resorting to the direct exchange of information among individuals.

Bee Spoke

REGARDLESS of the waggle dance's evolutionary origin, the idea that bees could possess a symbolic language has never been simply relegated to an incidental by-product of their flying existence, a serendipity of evolution. Anthropologists endlessly debate whether it is possible to have thinking without language, whether one has to be able to formulate thoughts into something resembling words to be truly sentient. And they take enormous pains to define what makes our language special, and how it can be the one holdout in making humans different from all other animals. But the bees make us ask instead whether it is possible to have language without thinking, since even the most ardent admirers of the waggle dance do not maintain that the bees' cognitive capacities mirror our own. So do the bees speak? And if so, does it mean we have to admit them into a special club, unlike any other animal?

Although many beekeepers had noticed that single foraging bees seem to advertise the location of nectar-rich patches of flowers to the rest of the hive, the first detailed description of the forager's performance was made in 1919 by the Austrian scientist Karl von Frisch, who shared the 1973 Nobel Prize in physiology and medicine with the ethologists Konrad Lorenz and Niko Tinbergen for his accomplishment. He was able to carefully track the movements of individual bees by placing his colonies in glass-walled observation hives and marking the bees with either dabs of paint or tiny numbered circles that he glued to their backs.

Von Frisch noted that when a worker bee returned to the hive after visiting a rich food source, she performed a stereotyped series of movements on the surface of the honeycomb. If the food source is close by, less than 50 yards or so, she did a rather simple "round dance," in which the forager runs in narrow circles. More distant food patches warranted a "waggle dance," which contains information about both the distance of the food from the hive and the direction in which it lies. The waggle dance consists of a straight run followed by a semicircle first to one side and then another, in a rough figure eight, with the bee waggling her abdomen energetically during the straight run.

The length of the run is correlated with the distance of the food from the hive, while the angle of the bee's body relative to vertical indicates the angle between the sun and the food source. The vibrating wings of the dancing bee also convey auditory information to the rest of the hive; silenced bees do not recruit others to the food source, and it makes sense that sound would be needed, since the inside of the hive is dark and the other workers cannot simply watch what the dancer is doing. Once the dancer has completed her performance, other bees venture out of the hive and go, more or less directly, to the location she indicated.

In other words, the bees seem to have symbolic representations for the distance and direction of the food, which fits many if not all of the criteria for an actual language. This was big news. Historian of science Tania Munz points out that during the 1960s, bee language was "the most widely studied form of animal communication and some deemed it the most complex second only to human speech." Even Carl Jung took note, musing that we would interpret the bees' behavior, if it occurred in humans, "as a conscious and intentional act and can hardly imagine how anyone could prove in a court of law that it had taken place unconsciously. . . . Nor is there any proof that bees are unconscious." Those with a yearning to see the waggle dance for themselves need look no further than YouTube, of course; one video of a dancing bee had nearly eighty thousand hits, and enthusiastic if sometimes inadvertently ironic comments such as, "I couldn't do that. Bees are smarter than me,"

"Why would you shake your butt as communication, weird," or, even better, "Wow. Their [sic] smart."

Although von Frisch's discovery was mainly greeted with amazement and rather uncritical acceptance by both the general public and other scientists, a few remained skeptical that the bees were truly capable of using the sophisticated information encoded in the dances. Foremost among these was Adrian Wenner, a professor at the University of California, Santa Barbara—and, in the spirit of full disclosure, my former teacher and mentor as an undergraduate. A soft-spoken but determined man, Wenner did not dispute the information contained in the waggle dance; he could observe a returning forager and calculate the distance and direction of the food patch perfectly well himself. He just didn't think the bees were using the information.

Wenner claimed that a much simpler explanation for how the bees found the food existed: the other workers simply smelled the odor that lingered on the recruiter's body, left the hive and flew, sniffing the air, until they perceived the same scent emanating from a patch of flowers. The experiments that von Frisch and other scientists performed demonstrated merely that the bees found the food, he said, not how they did so. His hypothesis was much more parsimonious, and hence, Wenner concluded, scientists were obliged to use it rather than the more elaborate explanation that required talking bees. Why, then, did the bees dance, and why did the dance contain information that was interpretable by humans, if the bees didn't use it? Wenner would always smile an impish smile when asked that question, and point out that nature did not evolve for a purpose—to suggest that it did was teleological and unscientific. The dance didn't have to be used by the bees in the way we could use it; a cricket's call can be used to calculate the temperature because he sings more quickly when it is warmer, but no one has ever suggested that the crickets evolved their chirps so that they could act as thermometers.

Wenner's iconoclastic views were not particularly popular, which he also attributed to an unscientific bias toward wanting to believe the more dramatic and exciting story of an insect language. Eventually, however, scientists began to pit the two ideas against each other. To some extent, it is unfair to claim that von Frisch dismissed the use of odor cues by the bees, since he did acknowledge its role in some of his papers, and indeed most researchers acknowledge that the bees in the hive do not ignore the information contained in the scent of a returning worker.

Many biologists were convinced that the bees do indeed use the information in the dance by some experiments published in 1975 by James L. Gould, in which he manipulated the dancer to "lie" about where the food was located using a flashlight to mimic the sun and, hence, alter the angle at which the dance was produced relative to the sun's actual position. Wenner was unconvinced, suggesting that the experiment was never replicated, and he and a few other scientists also claimed that Gould did not fully control for the bees' use of odors as an alternative explanation.

Several scientists have tried to manufacture artificial bees that could be made to dance inside the hive to further test the hypotheses, and one of these was able to recruit at least some bees to the food source it was programmed to dance about to the rest of the colony. Wenner once again dismissed these findings as inconclusive, and it is certainly the case that the mechanical bees didn't do the job nearly as well as a real one.

The conclusive set of experiments, at least in the majority of scientists' view, came from H. Esch and colleagues, who were able to manipulate something called the *optic flow* perceived by the bees. Bees measure distance by gauging the way images in the environment move across their eyes as they fly, rather like clocking the trees that tick by the windows of a moving train. The scientists trained the bees to fly through a tunnel lined with a black and white pattern that presented an optical illusion to the insects, making it appear that they had flown a longer distance than they actually had. When the fooled bees got back to the hive, they produced a dance that indicated the food was farther away than it was. The recruits promptly flew to the wrong site, indicating that they had indeed been misled by the dance itself.

Yet other studies used harmonic radar to track individual bees and the flight paths they took to the feeder or flower patch; these showed that most of the bees recruited by a dancer took a straight path to the food, rather than zigzagging back and forth the way they would be expected to if they were simply using the odors in the air to find the patch that smelled like the dancer inside the hive.

Finally, my friend Kirk Visscher and a former student of his, Gavin Sherman, demonstrated that the waggle dances help the bees survive in nature. In a clever experiment, they used lighting to mimic the sun and misdirect the bees, so that the dances didn't help the members of the hive to find a food source. They allowed a control set of colonies to dance appropriately. At the end of the season, the deceived colonies had accumulated significantly less honey than those in the control group, an important consideration in the well-being of the hive.

No one, including Kirk, disputes that the bees also use the odor information from the initial dancer to find food, and that under some circumstances the dances are not needed for the colony members to find out where they can forage. At some times of year, the bees can find plenty of nectar by relying simply on the flower odors in their environment that are carried on the bodies of the workers that find each patch of blossoms. But when the going gets tough, the bees seem to need dancing. Some theoretical work by Madeleine Beekman and Jie Bin Lew of the University of Sydney in Australia formalized this mathematically, demonstrating that dancing helps a colony concentrate on the best food sources in the area and not waste time sending workers to a low-quality patch. It is most helpful when the probability of finding a patch on one's own as an independent forager is relatively low, because the dance allows the colony to exploit only the richest nectar sources.

Bilingual Bees?

WHERE did the bee dance language come from? I suggested earlier that scientists believe it may have evolved from the need to hide signals of food sources from competitors, but that problem is common to many group-living animals, not just honeybees. Which species have dance language, and how it is used, sheds light on the origin of this behavior.

Beekman and her colleagues study dance language in the red dwarf bee of southeast Asia, which is closely related to the honeybee but nests in the open, making a single comb that is suspended from a shaded branch, rather like a wasp nest. This nesting habit is thought to be more like the ancestral state, with the elaborate cavity nests of the honeybees and several other species being more recently evolved. Dwarf bees still dance, and like honeybees, they do so both when they are selecting a new nest site and when they are feeding. But giving directions to a new nest site on a branch out in the open poses a much different problem than telling the rest of the colony how to find a small entrance hole to a cavity, such as the honeybees use. It's like the difference between telling someone to head north on State Street until he sees a big pink building—"you can't miss it"—versus giving explicit directions to a door on the third floor, east wing, of that building. On the other hand, a flower patch is a flower patch, and directions don't need to be all that precise when a worker is dancing to indicate where her nest mates can find food.

By videotaping dances of the dwarf bees, Beekman and her colleagues found that the dances used for both food and nest site directions were equally imprecise. Honeybees, in contrast, are far sloppier when they dance to show the other workers where food is than when they are directing them to a new home. The scientists believe that the dance evolved as a way to convey information about the new nest site, and that its use to indicate food sources came later.

If more than one species of bee uses symbolic language, can they understand each other? In a paper titled "East Learns from West," Songkun Su from Zheijiang University in China and coworkers showed that Asian honeybees, which are a different species than the European honeybees commonly found in Europe and the species that has been introduced to North America, can follow directions given by their European counterparts. Su and colleagues painstakingly constructed colonies containing a queen of one species and workers of the other, a daunting task because the specific odors of each colony usually mean the different species detect and kill any outsiders. Ordinarily the dances from the two bee species differ in what might be called dialect, with variations in the duration of the waggle portion of the dance. Su demonstrated that the two species could follow each other's directions, which means that the bees must learn some elements of the dance language.

Bees, Chimps, and Symbols

DESPITE the great interest in bee communication and the ever-greater elucidation of the dance, as Wenner points out, no one has ever been able to use the information to direct bees to particular crops that need pollinating or to sources of nectar that would be preferred by humans for honey production. So what is the significance of the bee dance language?

As I have mentioned before, we seem almost obsessed with setting out criteria for membership in a club that only we can enter; humans are the only species to use tools, for example, or to routinely kill members of our own species without using them for food. Both of these turn out to be unwarranted—chimpanzees, crows, and several other animals use tools, and fig wasps, among other species, routinely slaughter their own. One can detect a certain desperation in resorting to homicidal violence as a badge of distinction, but the effort continues. And language, with its slang and poetry, has always remained a prime candidate.

The problem is that many if not most other animals communicate, too, and they communicate in often complex and sophisticated ways. As Alison Wray put it in a book titled *Language Origins,* "Pinning down precisely what it is that makes human language special has never been so difficult. It's not that we no longer regard it as special, but almost every time you think you have a feature that helps define the real essence of language, or that provides a necessary context for its emergence, you seem to find some other animal that does it as well."

When that other animal is an insect, the comparison seems particularly troubling. Eileen Crist, in an analysis of the bee language controversy, says, "This almost-serious idea of an insect with language has had an unsettling effect in behavioral science." She notes that the waggle dance satisfies the criteria of having a set of rules, with a necessary order and complexity of the symbols that are used. Psychologists Mark Hauser, Noam Chomsky, and Tecumseh Fitch declare that human language is qualitatively different from other forms of animal communication, whether birdsong or bee dances, at least if one distinguishes between what they call the faculty of language in the broad sense and the narrow sense; it isn't clear whether bees get to join at least one of those circles. Other language scholars struggle with the distinction, sometimes mentioning the auditory sensitivity that enables the nearly endless discrimination among different sounds. The idea that Washoe the chimp or Alex the African grey parrot could be taught elements of our own language makes us reexamine our uniqueness yet again.

Just as an aside, amid all the hand-wringing and contention about whether what the bees do is really "language," no one seems to question whether it's really "dance." Maybe the dance scholars are just more easygoing than the linguists, or maybe we are already comfortable sharing that capacity with other species, though one could argue that the struts and tail shimmies of a peacock are hardly analogous to a waltz. But this points to the futility of the discussion; if we always narrow our definition of language, sooner or later we will end up with a capacity only we can possess. The breathtaking displays of a bird of paradise, or the comical movements of a lizard extending its dewlap, do not detract from the achievements of a ballerina.

It seems to me that the bees are not much like Alex or Washoe, because we can't teach them to say "cup," or to comment on the day's activities, or ask for another piece of fruit. Bees only talk about what they need to, mainly involving food or a place to live, and I think we have gotten way too interested in the accident of their using representational movements in communicating those objectives. Bee language didn't arise from a common ancestor with humans, which means we can't see it as a primitive version of our own language. This forces us to be less anthropomorphic than we are with the primates. And the less anthropomorphic we are, the more incredible the bees' accomplishment becomes, because they evolved this system of communication with entirely different selection pressures than the ones that led to human language. How did evolution take such different paths to get to superficially similar outcomes?

If it's true that bees needed a way to hide their communication from rivals, and that the ants' method of hauling colony members off to the new nest was unworkable for a flying insect, two other questions remain that in my mind are much more interesting than endless fussing over who does and doesn't qualify to enter the human club. The first is why all other social flying insects did not evolve some version of dance language. The second is why, when we are much more like the ants than the air-bound bees, we humans evolved language ourselves, rather than just dragging each other around when we wanted to convey a decision. Maybe human language isn't unique. But it beats at least one of the obvious alternatives.

Bee language, and the complicated decisions that accompany it, exemplifies why we keep coming back to insects, why, despite their encroachment on our kitchens and sometimes our health, we can't shake our simultaneous sense of connection and distance. We all want to be able to talk to animals. With bees, as with other insects, we can be pretty sure that we will never be able to communicate with them, even to the limited extent we can communicate with our pets. Fanciful interpretations such as T. H. White's and Maeterlinck's aside, no one really believes we could get a sense of what a bee is feeling, if it has feelings at all. And yet although we can't talk to them, they seem to be able to talk to each other, in a way that can be said to be more sophisticated than the chirps and buzzes common throughout the rest of the animal kingdom. It is precisely this sense of them being more like us than anything else, with their elaborate houses, facial recognition, use of others' labor, and complicated symbols, and yet so impossibly different from the inside out, that keeps us hooked.

Nearly a century ago, as the Jazz Age gathered steam, Don Marquis was a writer for the *Evening Sun* in New York. His work encompassed many topics, but he is best remembered for creating—or at least transcribing—archy, a cockroach that wrote free verse on Marquis's typewriter by laboriously crashing his head into the keys. Archy's inability to hold down the shift key led to all of his poems being written in lower case, which added a certain insouciance to his observations. The poems were first collected in 1927 and were followed by several additional volumes, with a new trove of Marquis's work discovered in 1986. The poems proved remarkably popular, coming to include commentary by the insect's friend mehitabel, "an alley cat of questionable character," as she is described on DonMarquis.com.

Archy was generally blunt about the human condition, but he reserved some of his most trenchant observations for the role of the six-legged in the lives of the bipedal. As I have throughout this book, archy questioned the careless hubris of our assumption of superiority:

> men talk of money and industry
> of hard times and recoveries
> of finance and economics
> but the ants wait and the scorpions wait
> for while men talk they are making deserts all the time
> getting the world ready for the conquering ant
> drought and erosion and desert
> because men cannot learn

The descendents of archy and his kind no longer seem to be leaving us missives, which seems rather a shame. Perhaps our modern keyboards seem daunting, with their need to be connected to vast processors that are inaccessible to a mere cockroach lacking both strength to turn the switch as well as a password. I would give a great deal to come into my office one day and find something like this, one of archy's best, on my monitor:

> i do not see why men
> should be so proud
> insects have the more
> ancient lineage
> according to the scientists
> insects were insects
> when man was only
> a burbling whatisit

One can only hope that perhaps one day soon our modern cockroaches will learn to manipulate a touchpad.

ACKNOWLEDGMENTS

My first thanks are owed to Bill Cade, orthopterist extraordinaire, who in addition to providing much help with my own cricket research over the years is the originator of the title *Sex on Six Legs.* He graciously allowed me to use it, though I am sure that his interpretation of the topic would have been at least as compelling. Other colleagues were generous with their unpublished data or manuscripts, anecdotes, and careful reading of several of the chapters: particular thanks to Nathan Bailey, Dave Featherstone, Ryan Gregory, Joan Herbers, and Kirk Visscher. Kirk has always been a great source of bee lore and other social insect information, not to mention supplying us with honey and helping us get rid of our personal swarm. Leigh Simmons has continued to be a great collaborator and colleague. Much of my love of insects as study organisms came from Adrian Wenner, who also taught me a great deal about the practice of science and its potential pitfalls. My Ph.D. advisor, the late W. D. Hamilton, was a key influence in my thinking about evolutionary biology, as well as a source of appreciation for the wonders of insect natural history. My students, graduate and undergraduate, always provide interesting comments and acted as sounding boards for many of the ideas presented here.

My agent Wendy Strothman has been a steady support for my writing, and no one could ask for a better editor than Andrea Schulz, even if she did keep suggesting that the book contain "less sex" when I was working on the proposal. I am also grateful to several "real" science writers who have given me advice and encouragement about writing for the public: Deborah Blum, Virginia Morell, and Carl Zimmer. Finally, John Rotenberry (mostly) graciously suffers being dragged along on entomological endeavors, despite being a "bird person," and has supported me in this as in all of my efforts.

REFERENCES

Introduction: Life on Six Legs

Darwin, C. 2001. *The Voyage of the Beagle.* Reprint. New York: Modern Library.

Dawkins, R. 2005. Introduction: The illusion of design. *Natural History,* November.

Dethier, V. G. 1964. Microscopic brains. *Science* 143: 1138–1145.

———. 1981. Fly, rat, and man: The continuing quest for an understanding of behavior. *Proceedings of the American Philosophical Society* 125: 460–466.

Gal, R., and F. Libersat. 2010. A wasp manipulates neuronal activity in the sub-esophageal ganglion to decrease the drive for walking in its cockroach prey. *PLoS ONE* 5: E10019.

Gallai, N., J. M. Salles, J. Settele, and B. E. Vaissiere. Economic valuation of the vulnerability of world agriculture confronted with pollinator decline. *Ecological Economics* 68: 810–821.

Hoyt, E., and T. Schultz, eds. 1999. *Insect Lives.* New York: John Wiley and Sons.

Losey, J. E., and M. Vaughan. 2006. The economic value of ecological services provided by insects. *BioScience* 56: 311–323.

Ratnieks, F. L. W. 2006. Can humans learn from insect societies? *Nova Acta Leopoldina NF* 93: 97–116.

Siebert, C. 2009. Something wild. *New York Times,* March 5.

Vosshall, L. B. 2007. Into the mind of a fly. *Nature* 450: 193–197.

Wallechinsky, D., I. Wallace, and A. Wallace. 1977. *The Book of Lists.* New York: William Morrow and Co.

Zimmer, C. 2010. A wasp finds the seat of the cockroach soul. *Discover Blogs, The Loom.* April 20. Available at http://blogs.discovermagazine.com/loom/2010/04/20/a-wasp-finds-the-seat-of-the-cockroach-soul/.

1. If You're So Smart, Why Aren't You Rich?

Burger, J. M. S., M. Kolss, J. Pont, and T. J. Kawecki. 2008. Learning ability and longevity: A symmetrical evolutionary trade-off in *Drosophila. Evolution* 62: 1294–1304.

Chanda, S., and E. Caulton. 1999. David Douglas Cunninghan (1843–1914): A biographical profile. *Aerobiologia* 15: 255–258.

Chittka, L., and E. Leadbeater. 2005. Social learning: Public information in insects. *Current Biology* 15: R869–R871.

Clare, S. 2006. Honeybees make plans. *Journal of Experimental Biology* 209: ii.

Collett, T. S. 2008. Insect behaviour: Learning for the future. *Current Biology* 18: R131–R134.

Csibra, G. 2007. Teachers in the wild. *Trends in Cognitive Sciences* 11: 95–96.

Cunningham, D. D. 1907. *The Plagues and Pleasures of Life in Bengal.* London: John Murray.

Dacke, M., and M. V. Srinivasan. 2008. Evidence for counting in insects. *Animal Cognition* 11: 683–689.

D'Ettorre, P. 2007. Evolution of sociality: You are what you learn. *Current Biology* 17: R766–R768.

Dukas, R. 2008. Evolutionary biology of insect learning. *Annual Review of Entomology* 53: 145–160.

Dukas, R., C. Clark, and K. Abbott. 2006. Courtship strategies of male insects: When is learning advantageous? *Animal Behaviour* 72: 1395–1404.

Dyer, A. G., C. Neumeyer, and L. Chittka. 2005. Honeybee (*Apis mellifera*) vision can discriminate between and recognise images of human faces. *Journal of Experimental Biology* 208: 4709–4714.

Dyer, A. G., M. G. P. Rosa, and D. H. Reser. 2008. Honeybees can recognise images of complex natural scenes for use as potential landmarks. *Journal of Experimental Biology* 211: 1180–1186.

Fabre, J. H. 1981. *The Insect World of J. Henri Fabre.* Reprint. New York: Harper Colophon.

Fisher, O. 1911. Insect intelligence. *Nature* 86: 144.

Kennerknecht, I., N. Pluempe, and B. Welling. 2008. Congenital prosopagnosia — A common hereditary cognitive dysfunction in humans. *Frontiers in Bioscience* 13: 3150–3158.

Kosmos, H. 2008. Through the eyes of a bee. Interview with Adrian Dyer. *Humboldt Kosmos.* Available at www.humboldt-foundation.de/web/kosmos-interviews-en-91-1.html.

Leadbeater, E., and L. Chittka. 2007. Social learning in insects — From miniature brains to consensus building. *Current Biology* 17: R703–R713.

Leadbeater, E., N. E. Raine, and L. Chittka. Social learning: Ants and the meaning of teaching. *Current Biology* 16: R323–R325.

Mery, F., A. T. Belay, A. K.-C. So, M. B. Sokolowski, and T. J. Kawecki. 2007. Natural polymorphism affecting learning and memory in *Drosophila. Proceedings of the National Academy of Sciences USA* 104: 13051–13055.

Mery, F., and T. J. Kawecki. 2004. The effect of learning on experimental evolution of resource preference in *Drosophila melanogaster. Evolution* 58: 757–767.

———. 2005. A cost of long-term memory in *Drosophila. Science* 308: 1148.

Nowbahari, E., A. Scohier, J.-L. Durand, and K. L. Hollis. 2009. Ants, *Cataglyphis cursor,* use precisely directed rescue behavior to free entrapped relatives. *PLoS ONE* 4: E6573.

Paenke, I., B. Sendhoff, and T. J. Kawecki. 2007. Influence of plasticity and learning on evolution under directional selection. *American Naturalist* 170: E47–E58.

Richardson, T. O., P. A. Sieeman, J. M. McNamara, A. I. Houston, and N. R. Franks. 2007. Teaching with evaluation in ants. *Current Biology* 17: 1520–1526.

Sitaraman, D., M. Zars, H. LaFerriere, Y.-C. Chen, A. Sable-Smith, T. Kitamoto, G. E. Rottinghaus, and T. Zars. 2008. Serotonin is necessary for place memory in *Drosophila. Proceedings of the National Academy of Sciences USA* 105: 5579–5584.

Thornton, A. 2008. Variation in contributions to teaching by meerkats. *Proceedings of the Royal Society of London B* 275: 1745–1751.

Thornton, A., N. J. Raihani, and A. N. Radford. 2007. Teachers in the wild: Some clarification. *Trends in Cognitive Sciences* 11: 272–273.

Tomchik, S. M., and R. Davis. 2008. Out of sight, but not out of mind. *Nature* 453: 1192–1194.

Wessnitzer, J., M. Mangan, and B. Webb. 2008. Place memory in crickets. *Proceedings of the Royal Society of London B* 275: 915–921.

Zhang, S., S. Schwarz, M. Pahl, H. Zhu, and J. Tautz. 2006. Honeybee memory: A honeybee knows what to do and when. *Journal of Experimental Biology* 209: 4420–4428.

Zimmer, C. 2008. Lots of animals learn, but smarter isn't better. *New York Times,* May 6.

2. Six Legs and a Genome

Birney, E. 2007. Come fly with us. *Nature* 450: 184–185.

Brenner, S. 1996. Interview: The world of genome projects. *BioEssays* 12: 1039–1042.

Chadee, D. D., P. Kittayapong, A. C. Morrison, and W. J. Tabachnick. 2007. A breakthrough for global public health. *Science* 316: 1703–1704.

Check, E. 2006. From hive minds to humans. *Nature* 443: 893.

Cusson, M. 2008. The molecular biology toolbox and its use in basic and applied insect science. *BioScience* 58: 691–700.

Evans, J. D., and D. Gundersen-Rindal. 2003. Beenomes to *Bombyx:* Future directions in applied insect genomics. *Genome Biology* 4: 107.

Flannery, M. C. 2007. One genome, one piece of the puzzle. *American Biology Teacher* 69: 109–112.

———. 2008. Insects by the numbers. *American Biology Teacher* 70: 426–429.

Gregory, T. R. 2005. Synergy between sequence and size in large-scale genomics. *Nature Reviews Genetics* 6: 699–708.

———. 2005. Genome size evolution in animals. In *The Evolution of the Genome,* ed. T. R. Gregory. New York: Elsevier.

Gregory, T. R., and J. S. Johnston. 2008. Genome size diversity in the family Drosophilidae. *Heredity* 101: 228–238.

Gunter, C. 2007. Genomics on the fly. *Nature Reviews Genetics* 8: 904.

Jenner, R. A., and M. A. Wills. 2007. The choice of model organisms in evo-devo. *Nature Reviews Genetics* 8: 311–319.

Koshikawa, S., S. Miyazaki, R. Cornette, T. Matsumoto, and T. Miura. 2008. Genome size of termites (Insecta, Dictyoptera, Isoptera) and wood roaches (Insecta, Dictyoptera, Cryptocercidae). *Naturwissenschaften* 95: 859–867.

Ledford, H. 2007. Attack of the genomes. *Nature* 450: 142–143.

Maderspacher, F. 2008. Genomics: An inordinate fondness for beetles. *Current Biology* 18: R466.

Myrmecos blog. 2009. Which ants should we target for genome sequencing? January 15. Available at http://myrmecos.net/.

National Human Genome Research Institute. 2010. NHGRI website. Available at http://genome.gov.

Pennisi, E. 2007. Fruit fly blitz shows the power of comparative genomics. *Science* 318: 903.

Ponting, C. P. 2008. The functional repertoires of metazoan genomes. *Nature Reviews Genetics* 9: 689–698.

Robinson, G. E., and Y. Ben-Shahar. 2002. Social behavior and comparative genomics: New genes or new gene regulation? *Genes, Brain, and Behavior* 1: 197–203.

Smith, C. R., A. L. Toth, A. V. Suarez, and G. E. Robinson. 2008. Genetic and genomic analyses of the division of labour in insect societies. *Nature Reviews Genetics* 9: 735–748.

Thompson, G. J., H. Yockey, J. Lim, and B. P. Oldroyd. 2007. Experimental manipulation of ovary activation and gene expression in honey bee (*Apis mellifera*) queens and workers: Testing hypotheses of reproductive regulation. *Journal of Experimental Zoology* 307A: 600–610.

Toth, A. L., and G. E. Robinson. 2007. Evo-devo and the evolution of social behavior. *Trends in Genetics* 23: 334–341.

Toth, A. L., K. Varala, T. C. Newman, F. E. Miguez, S. K. Hutchison, D. A. Willoughby, J. F. Simons, M. Egholm, J. H. Hunt, M. E. Hudson, and G. E. Robinson. 2007. Wasp gene expression supports an evolutionary link between maternal behavior and eusociality. *Science* 318: 441–444.

Tribolium Genome Sequencing Consortium. 2008. The genome of the model beetle and pest *Tribolium castaneum*. *Nature* 452: 949–955.

Tsutsui, N. D., A. V. Suarez, J. C. Spagna, and J. S. Johnston. 2008. The evolution of genome size in ants. *BMC Evolutionary Biology* 8: 64.

Wade, N. 2000. Scientist at work: Sydney Brenner. *New York Times*, March 7.

Waterhouse, R. M., S. Wyder, and E. M. Zdobnov. 2008. The *Aedes aegypti* genome: A comparative perspective. *Insect Molecular Biology* 17: 1–8.

Whitfield, J. 2007. Who's the queen? Ask the genes. *Science* 318: 910–911.

Wilson, E. O. 2006. How to make a social insect. *Nature* 443: 919–920.

Zagorski, N. 2006. Profile of Gene E. Robinson. *Proceedings of the National Academy of Sciences USA* 103: 16065–16067.

Zdobnov, E. M., and P. Bork. 2006. Quantification of insect genome divergence. *Trends in Genetics* 23: 16–20.

3. The Inner Lives of Wasps

Bell, A. M. 2007. Animal personalities. *Nature* 447: 539–540.

Biro, P. A., and J. A. Stamps. 2008. Are animal personality traits linked to life-history productivity? *Trends in Ecology and Evolution* 23: 361–368.

Cervo, R., L. Dapporto, L. Beani, J. E. Strassmann, and S. Turillazzi. 2008. On status badges and quality signals in the paper wasp *Polistes dominulus*: Body size, facial colour patterns and hierarchical rank. *Proceedings of the Royal Society of London B* 275: 1189–1196.

Darwin, C. 2009. *The Expression of the Emotions in Man and Animals*. Reprint of 1872 edition. New York: Penguin Classics.

Dethier, V. G. 1964. Microscopic brains. *Science* 143: 1138–1145.

D'Ettorre, P., and J. Heinze. 2005. Individual recognition in ant queens. *Current Biology* 15: 2170–2174.

Dreier, S., J. S. van Zweden, and P. D'Ettorre. 2007. Long-term memory of individual identity in ant queens. *Biology Letters* 3: 459–462.

Gosling, S. D. 2001. From mice to men: What can we learn about personality from animal research? *Psychological Bulletin* 127: 45–86.

Gosling, S. D., and S. Vazire. 2002. Are we barking up the right tree? Evaluating a comparative approach to personality. *Journal of Research in Personality* 36: 607–614.

Griffin, D. R. 1984. *Animal Thinking*. Cambridge, MA: Harvard University Press.

———. 2001. *Animal Minds*. Chicago: University of Chicago Press.

Gronenberg, W., L. E. Ash, and E. A. Tibbetts. 2008. Correlation between facial pattern recognition and brain composition in paper wasps. *Brain, Behavior, and Evolution* 71: 1–14.

Higgins, L. A., K. M. Jones, and M. L. Wayne. 2005. Quantitative genetics of natural variation of behavior in *Drosophila melanogaster*: The possible role of the social environment on creating persistent patterns of group activity. *Evolution* 59: 1529–1539.

Keller, E. F. 1983. *A Feeling for the Organism*. New York: W. H. Freeman and Co.

Koolhaas, J. M. 2008. Coping style and immunity in animals: Making sense of individual variation. *Brain, Behavior, and Immunity* 22: 662–667.

Kortet, R., and A. Hedrick. 2007. A behavioural syndrome in the field cricket *Gryllus integer*: Intrasexual aggression is correlated with activity in a novel environment. *Biological Journal of the Linnean Society* 91: 475–482.

Mehta, P. H., and S. D. Gosling. 2008. Bridging human and animal research: A comparative approach to studies of personality and health. *Brain, Behavior, and Immunity* 22: 651–661.

Nemiroff, L., and E. Despland. 2007. Consistent individual differences in the foraging behaviour of forest tent caterpillars (*Malacosoma disstria*). *Canadian Journal of Zoology* 85: 1117–1124.

Nettle, D. 2006. The evolution of personality variation in humans and other animals. *American Psychologist* 61: 622–631.

Øyvind, Ø. 2007. Preface: Plasticity and diversity in behavior and brain function—Important raw material for natural selection? *Brain, Behavior, and Evolution* 70: 215–217.

Réale, D., S. M. Reader, D. Sol, P. T. McDougall, and N. J. Dingemanse. 2007. Integrating animal temperament within ecology and evolution. *Biological Reviews* 82: 291–318.

Robins, R. W. 2005. The nature of personality: Genes, culture, and national character. *Science* 310: 62–63.

Sih, A., A. M. Bell, and J. C. Johnson. 2004. Behavioral syndromes: An integrative overview. *Quarterly Review of Biology* 79: 241–277.

———. 2004. Behavioral syndromes: An ecological and evolutionary overview. *Trends in Ecology and Evolution* 19: 372–378.

Sih, A., and J. V. Watters. 2005. The mix matters: Behavioural types and group dynamics in water striders. *Behaviour* 142: 1417–1431.

Stamps, J. A. 2007. Growth-mortality tradeoffs and "personality traits" in animals. *Ecology Letters* 10: 355–363.

Tibbetts, E. A. 2002. Visual signals of individual identity in the wasp *Polistes fuscatus. Proceedings of the Royal Society of London B* 269: 1423–1428.

———. 2004. Complex social behaviour can select for variability in visual features: A case study in *Polistes* wasps. *Proceedings of the Royal Society of London B* 271: 1955–1960.

Tibbetts, E. A., and J. Dale. 2004. A socially enforced signal of quality in a paper wasp. *Nature* 432: 218–222.

———. 2007. Individual recognition: It is good to be different. *Trends in Ecology and Evolution* 22: 529–537.

Tibbetts, E. A., and R. Lindsay. 2008. Visual signals of status and rival assessment in *Polistes dominulus* paper wasps. *Biology Letters* 4: 237–239.

Wilson, D. S., A. B. Clark, K. Coleman, and T. Dearstyne. 1994. Shyness and boldness in humans and other animals. *Trends in Ecology and Evolution* 9: 442–446.

Wolf, M., G. S. van Doorn, O. Leimar, and F. J. Weissing. 2007. Life-history trade-offs favour the evolution of animal personalities. *Nature* 447: 581–584.

4. Seinfeld and the Queen

Angier, N. 2007. In Hollywood hives, the males rule. *New York Times,* November 13.

Brackney, S. 2007. The real life of bees. *New York Times,* November 9.

Charlat, S., E. A. Hornett, J. H. Fullard, N. Davies, G. K. Roderick, N. Wedell, and G. D. D. Hurst. Extraordinary flux in sex ratio. *Science* 317: 214.

Cobb, M. 2002. Jan Swammerdam on social insects: A view from the seventeenth century. *Insectes Sociaux* 49: 92–97.

Crane, E. 1999. *The World History of Beekeeping and Honey Hunting.* New York: Routledge.
Godfray, H. C. J., and J. H. Werren. 1996. Recent developments in sex ratio studies. *Trends in Ecology and Evolution* 11: 59–63.
Hamilton, W. D. 1967. Extraordinary sex ratios. *Science* 156: 477–488.
Swammerdam, J. 2004. Information available at http://janswammerdam.net.
Trivers, R. L., and H. Hare. 1976. Haplodiploidy and the evolution of the social insect. *Science* 191: 249–263.
Trivers, R. L., and D. E. Willard. 1973. Natural selection of parental ability to vary the sex ratio of offspring. *Science* 179: 90–92.
Wilson, B. 2004. *The Hive: The Story of the Honeybee and Us.* London: John Murray.
Zuk, M. 2002. *Sexual Selections: What We Can and Can't Learn about Sex from Animals.* Berkeley: University of California Press.

5. Sperm and Eggs on Six Legs

Ben-Ari, E. T. 2000. Choosy females. *BioScience* 50: 7–12.
Birkhead, T. R. 2000. Defining and demonstrating postcopulatory female choice — Again. *Evolution* 54: 1057–1060.
Birkhead, T. R., and T. Pizzari. 2002. Postcopulatory sexual selection. *Nature Reviews Genetics* 3: 262–273.
Bjork, A., Dallai, R., and S. Pitnick. 2007. Adaptive modulation of sperm production rate in *Drosophila bifurca,* a species with giant sperm. *Biology Letters* 3: 517–519.
Briceño, R. D., W. G. Eberhard, and A. S. Robinson. 2007. Copulation behaviour of *Glossina pallidipes* (Diptera: Muscidae) outside and inside the female, with a discussion of genitalic evolution. *Bulletin of Entomological Research* 97: 471–488.
Chapman, T. 2008. The soup in my fly: Evolution, form and function of seminal fluid proteins. *PLoS Biology* 6: 1379–1382.
Córdoba-Aguilar, A. 2006. Sperm ejection as a possible cryptic female choice mechanism in Odonata (Insecta). *Physiological Entomology* 31: 146–153.
Eberhard, W. G. 1991. Copulatory courtship and cryptic female choice in insects. *Biological Reviews* 66: 1–31.
Eberhard, W. G., and C. Cordero. 1995. Sexual selection by cryptic female choice on male seminal products — A new bridge between sexual selection and reproductive physiology. *Trends in Ecology and Evolution* 10: 493–496.

Engqvist, L. 2007. Nuptial gift consumption influences female remating in a scorpionfly: Male or female control of mating rate? *Evolutionary Ecology* 21: 49–61.

Fedina, T. Y. 2006. Cryptic female choice during spermatophore transfer in *Tribolium castaneum* (Coleoptera: Tenebrionidae). *Journal of Insect Physiology* 53: 93–98.

Holland, B., and W. R. Rice. 1997. Cryptic sexual selection — More control issues. *Evolution* 51: 321–324.

Holman, L., and R. R. Snook. 2008. A sterile sperm caste protects brother fertile sperm from female-mediated death in *Drosophila pseudoobscura. Current Biology* 18: 292–296.

Jagadeeshan, S., and R. S. Singh. 2006. A time-sequence functional analysis of mating behaviour and genital coupling in *Drosophila:* Role of cryptic female choice and male sex-drive in the evolution of male genitalia. *Journal of Evolutionary Biology* 19: 1058–1070.

Joly, D., C. Bressac, and D. Lachaise. 1995. Disentangling giant sperm. *Nature* 377: 202.

Kullmann, H., and K. P. Sauer. 2008. Mating tactic dependent sperm transfer rates in *Panorpa similis* (Mecoptera; Panorpidae): A case of female control? *Ecological Entomology* 34: 153–157.

LaMunyon, C. W., and T. Eisner. 1993. Postcopulatory sexual selection in an arctiid moth (*Utetheisa ornatrix*). *Proceedings of the National Academy of Sciences USA* 90: 4689–4692.

Martin, O., and M. Demont. 2008. Reproductive traits: Evidence for sexually selected sperm. *Current Biology* 18: R79–R81.

Miller, G. T., and S. Pitnick. 2002. Sperm-female coevolution in *Drosophila. Science* 298: 1230–1233.

Parker, G. 1970. Sperm competition and its evolutionary consequences in the insects. *Biological Reviews* 45: 525–567.

Pattarini, J. M., W. T. Starmer, A. Bjork, and S. Pitnick. 2006. Mechanisms underlying the sperm quality advantage in *Drosophila melanogaster. Evolution* 60: 2064–2080.

Peretti, A., W. G. Eberhard, and R. D. Briceño. 2006. Copulatory dialogue: Female spiders sing during copulation to influence male genitalic movements. *Animal Behavior* 72: 413–421.

Pitnick, S., G. S. Spicer, and T. A. Markow. 1995. How long is a giant sperm? *Nature* 375: 109.

Pizzari, T. 2006. Evolution: The paradox of sperm leviathans. *Current Biology* 16: R462–R464.

Simmons, L. W. 2001. *Sperm Competition and Its Evolutionary Consequences in the Insects.* Princeton, NJ: Princeton University Press.

———. 2005. The evolution of polyandry: Sperm competition, sperm selection, and offspring viability. *Annual Review of Ecology, Evolution, and Systematics* 36: 125–146.

Simmons, L. W., and F. Garcia-González. 2008. Evolutionary reduction in testes size and competitive fertilization success in response to the experimental removal of sexual selection in dung beetles. *Evolution* 62: 2580–2591.

Ward, P. I. 2000. Cryptic female choice in the yellow dung fly *Scathophaga stercoraria* (L.). *Evolution* 54: 1680–1686.

Wilson, N., S. C. Tubman, P. E. Eady, and G. W. Robertson. 1997. Female genotype affects male success in sperm competition. *Proceedings of the Royal Society of London B* 264: 1491–1495.

6. So Two Fruit Flies Go into a Bar...

Aldous, P. 2008. Randy flies reveal how booze affects inhibitions. *New Scientist,* January 3.

Bagemihl, B. 1999. *Biological Exuberance: Animal Homosexuality and Natural Diversity.* New York: St. Martin's Press.

Baram, M. 2007. If there was a gay-straight switch, would you switch? *ABC News,* December 14. Available at http://abcnews.go.com/Health/story?id=3997085&page=1.

Featherstone, D. E. 2010. Laboratory website. Available at www.uic.edu/depts/bios/faculty/featherstone/featherstone_d.shtml.

Gillespie, R. G. 1991. Homosexual mating behavior in male *Doryonychus raptor* (Araneae, Tetragnathidae). *Journal of Arachnology* 19: 229–230.

Grosjean, Y., M. Grillet, H. Augustin, J. F. Ferveur, and D. E. Featherstone. 2008. A glial amino-acid transporter controls synapse strength and courtship in *Drosophila. Nature Neuroscience* 11: 54–61.

Harari, A. R., H. J. Brockmann, and P. J. Landolt. 2000. Intrasexual mounting in the beetle *Diaprepes abbreviatus* (L.). *Proceedings of the Royal Society of London B* 267: 2071–2079.

Khamsi, R. 2005. Fruitflies tap in to their gay side. *Nature News,* June 2. Available at www.nature.com/news/2005/050531/full/news050531-9.html.

———. 2005. Gay flies lose their nerve. *BioEd Online,* November 9. Available at www.bioedonline.org/news/news.cfm?art=2153.

Kim, Y.-K., and L. Ehrman. 1998. Developmental isolation and subsequent adult behavior of *Drosophila paulistorum*. IV. Courtship. *Behavior Genetics* 28: 57–65.

Kimura, K., T. Hachiya, M. Koganezawa, T. Tazawa, and D. Yamamoto. 2008. Fruitless and doublesex coordinate to generate male-specific neurons that can initiate courtship. *Neuron* 59: 759–769.

Kyriacou, C. P. 2005. Sex in fruitflies is *fruitless*. *Nature* 436: 334–335.

Lee, H-G., Y.-C. Kim, J. S. Dunning, and K.-A. Han. 2008. Recurring ethanol exposure induces disinhibited courtship in *Drosophila*. *PLoS ONE* 3: E139.

Levan, K. E., T. Y. Fedina, and S. M. Lewis. 2008. Testing multiple hypotheses for the maintenance of male homosexual copulatory behaviour in flour beetles. *Journal of Evolutionary Biology* 22: 60–70.

LeVay, S. 1996. *Queer Science: The Use and Abuse of Research into Homosexuality*. Boston: MIT Press.

Liu, T., L. Dartevelle, C. Yuan, H. Wei, Y. Wang, J.-F. Ferveur, and A. Guo. 2008. Increased dopamine level enhances male-male courtship in *Drosophila*. *Journal of Neuroscience* 28: 5539–5546.

McRobert, S. P., and L. Tompkins. 1988. Two consequences of homosexual courtship performed by *Drosophila melanogaster* and *Drosophila affinis* males. *Evolution* 42: 1093–1097.

Miyamoto, T., and H. Amrein. 2008. Suppression of male courtship by a *Drosophila* pheromone inhibitor. *Nature Neuroscience* 11: 874–876.

Owen, J. 2005. Damselfly mating game turns some males gay. *National Geographic News*, June 21. Available at http://news.nationalgeographic.com/news/2005/06/0621_050622_gay_flies.html.

Preston-Mafham, K. 2006. Post-mounting courtship and the neutralizing of male competitors through "homosexual" mountings in the fly *Hydromyza livens* F. (Diptera: Scatophagidae). *Journal of Natural History* 40: 101–105.

Reinhardt, K., E. Harney, R. Naylor, S. Gorb, and M. T. Siva-Jothy. 2007. Female-limited polymorphism in the copulatory organ of a traumatically inseminating insect. *American Naturalist* 170: 931–935.

Rono, E., P. G. N. Njagi, M. O. Bashir, and A. Hassanali. 2007. Concentration-dependent parsiomonious releaser roles of gregarious male pheromone of the desert locust, *Schistocerca gregaria*. *Journal of Insect Physiology* 54: 162–168.

Serrano, J. M., L. Castro, M. A. Toro, and C. López-Fanjul. 1991. The genetic properties of homosexual copulation behavior in *Tribolium castaneum*: Diallel analysis. *Behavior Genetics* 21: 547–558.

Switzer, P. V., P. S. Forsythe, K. Escajeda, and K. C. Kruse. 2004. Effects of environmental and social conditions on homosexual pairing in the Japanese Beetle (*Popillia japonica* Newman). *Journal of Insect Behavior* 17: 1–16.

Tennent, W. J. 1987. A note on the apparent lowering of moral standards in the Lepidoptera. *Entomologist's Record* 99: 81–82.

Van Gossum, H., L. De Bruyn, and R. Stoks. 2005. Reversible switches between male-male and male-female mating behaviour by male damselflies. *Biology Letters* 1: 268–270.

Vosshall, L. B. 2008. Scent of a fly. *Neuron* 59: 685–689.

Wang, Q., L. Chen, J. Li, and X. Yin. 1996. Mating behavior of *Phytoecia rufiventris* Gautier (Coleoptera: Cerambycidae). *Journal of Insect Behavior* 9: 47–60.

7. Parenting and the Rotten Corpse

Beal, C. A., and D. W. Tallamy. 2006. A new record of amphisexual care in an insect with exclusive paternal care: *Rhynocoris tristis* (Heteroptera: Reduviidae). *Journal of Ethology* 24: 305–307.

Cocroft, R. 2002. Antipredator defense as a limited resource: Unequal predation risk in broods of an insect with maternal care. *Behavioral Ecology* 13: 125–133.

Costa, J. T. 2006. *The Other Insect Societies.* Cambridge, MA: Belknap Press.

Evans, T. A., E. J. Wallis, and M. A. Elgar. 1995. Making a meal of mother. *Nature* 376: 299.

Godfray, H.C.J. 1995. Evolutionary theory of parent-offspring conflict. *Nature* 376: 133–138.

———. 2005. Quick guide: Parent-offspring conflict. *Current Biology* 15: R191.

Goubault, M., D. Scott, and I. C. W. Hardy. 2007. The importance of offspring value: Maternal defence in parasitoid contests. *Animal Behaviour* 74: 437–446.

Klug, H., and M. B. Bonsall. 2007. When to care for, abandon, or eat your offspring: The evolution of parental care and filial cannibalism. *American Naturalist* 170: 886–901.

Kölliker, M. 2007. Benefits and costs of earwig (*Forficula auricularia*) family life. *Behavioral Ecology and Sociobiology* 61: 1489–1497.

Mas, F., and M. Kölliker. 2008. Maternal care and offspring begging in social insects: Chemical signalling, hormonal regulation and evolution. *Animal Behaviour* 76: 1121–1131.

Nakahira, T., and S. Kudo. 2008. Maternal care in the burrower bug *Adomerus triguttulus:* Defensive behavior. *Journal of Insect Behavior* 21: 306–316.

Ohba, S., K. Hidaka, and M. Sasaki. 2006. Notes on paternal care and sibling cannibalism in the giant water bug, *Lethocerus deyrolli* (Heteroptera: Belostomatidae). *Entomological Science* 9: 1–5.

Perry, J. C., and B. D. Roitberg. 2005. Ladybird mothers mitigate offspring starvation risk by laying trophic eggs. *Behavioral Ecology and Sociobiology* 58: 578–586.

———. 2006. Trophic egg laying: Hypotheses and tests. *Oikos* 112: 706–714.

Roy, H. E., H. Rudge, L. Goldrick, and D. Hawkins. 2007. Eat or be eaten: Prevalence and impact of egg cannibalism on two-spot ladybirds, *Adalia bipunctata. Entomologia Experimentalis et Applicata* 125: 33–38.

Santi, F., and S. Maini. 2007. Ladybirds mothers eating their eggs: Is it cannibalism? *Bulletin of Insectology* 60: 89–91.

Saul-Gershenz, L. S., and J. G. Millar. 2006. Phoretic nest parasites use sexual deception to obtain transport to their host's nest. *Proceedings of the National Academy of Sciences USA* 103: 14039–14044.

Smiseth, P. T., and H. J. Parker. 2008. Is there a cost to larval begging in the burying beetle *Nicrophorus vespilloides*? *Behavioral Ecology* 19: 1111–1115.

Smiseth, P. T., R.J.S. Ward, and A. J. Moore. 2006. Asynchronous hatching in *Nicrophorus vespilloides,* an insect in which parents provide food for their offspring. *Functional Ecology* 20: 151–156.

Smith, G., S. T. Trumbo, D. S. Sikes, M. P. Scott, and R. L. Smith. 2007. Host shift by the burying beetle, *Nicrophorus pustulatus,* a parasitoid of snake eggs. *Journal of Evolutionary Biology* 20: 2389–2399.

Smith, R. L. 1979. Paternity assurance and altered roles in the mating behaviour of a giant water bug, *Abedus herberti* (Heteroptera, Belostomatidae). *Animal Behaviour* 27: 716–725.

Staerkle, M., and M. Kölliker. 2008. Maternal food regurgitation to nymphs in earwigs (*Forficula auricularia*). *Ethology* 114: 844–850.

Steiger, S., K. Peschke, W. Francke, and J. K. Muller. 2007. The smell of parents: Breeding status influences cuticular hydrocarbon pattern in the burying beetle *Nicrophorus vespilloides. Proceedings of the Royal Society of London B* 274: 2211–2220.

Tallamy, D. W. 2005. Egg dumping in insects. *Annual Review of Entomology* 50: 347–370.

Tallamy, D. W., E. Walsh, and D. C. Peck. 2004. Revisiting parental care in the assassin bug, *Atopozelus pallens* (Heteroptera, Reduviidae). *Journal of Insect Behavior* 17: 431–436.

Tallamy, D. W., and T. K. Wood. 1986. Convergence patterns in subsocial insects. *Annual Review of Entomology* 31: 369–390.

Thomas, L. K., and A. Manica. 2003. Filial cannibalism in an assassin bug. *Animal Behaviour* 66: 205–210.

Trivers, R. L. 1974. Parent-offspring conflict. *American Zoologist* 14: 249–264.

Trumbo, S. T. 2006. Infanticide, sexual selection and task specialization in a biparental burying beetle. *Animal Behaviour* 72: 1159–1167.

———. 2007. Defending young biparentally: Female risk-taking with and without a male in the burying beetle, *Nicrophorus pustulatus*. *Behavioral Ecology and Sociobiology* 61: 1717–1723.

Williams, L., III, M. C. Coscarón, P. M. Dellapé, and T. M. Roane. 2005. The shield-backed bug, *Pachycoris stallii:* Description of immature stages, effect of maternal care on nymphs, and notes on life history. *Journal of Insect Science* 5: 1–13.

Zink, A. G. 2003. Quantifying the costs and benefits of parental care in female treehoppers. *Behavioral Ecology* 14: 687–693.

8. Pirates at the Picnic

Beebe, W. 1999. The hometown of the army ants. In *Insect Lives,* ed. E. Hoyt and T. Schultz. Reprint of 1921 edition. New York: John Wiley and Sons.

Beibl, J., R. J. Stuart, J. Heinze, and S. Foitzik. 2005. Six origins of slavery in formicoxenine ants. *Insectes Sociaux* 52: 291–297.

Bonckaert, W., K. Vuerinckx, J. Billen, R. L. Hammond, L. Keller, and T. Wenseleers. 2008. Worker policing in the German wasp *Vespula germanica*. *Behavioral Ecology* 19: 272–278.

Bono, J. M., M. F. Antolin, and J. M. Herbers. 2006. Parasite virulence and host resistance in a slave-making ant community. *Evolutionary Ecology Research* 8: 1117–1128.

Bono, J. M., E. R. Gordon, M. F. Antolin, and J. M. Herbers. 2006. Raiding activity of an obligate (*Polyergus breviceps*) and two facultative (*Formica puberula* and *F. gynocrates*) slave-making ants. *Journal of Insect Behavior* 19: 429–446.

Crompton, J. 1954. *Ways of the Ant.* Boston: Houghton Mifflin Co.

Foitzik, S., C. J. DeHeer, D. N. Hunjan, and J. M. Herbers. 2001. Coevolution in host-parasite systems: Behavioural strategies of slave-making ants and their hosts. *Proceedings of the Royal Society of London B* 268: 1139–1146.

Gadagkar, R. 2004. Why do honey bee workers destroy each other's eggs? *Journal of Bioscience* 29: 213–217.

Gloag, R., T. A. Heard, M. Beekman, and B. P. Oldroyd. 2008. Nest defence in a stingless bee: What causes fighting swarms in *Trigona carbonaria* (Hymenoptera, Meliponini)? *Insectes Sociaux* 55: 387–391.

Helanterä, H. 2007. How to test an inclusive fitness hypothesis — Worker reproduction and policing as an example. *Oikos* 116: 1782–1788.

Herbers, J. M. 2006. The loaded language of science. *Chronicle of Higher Education* 52: B5.

———. 2007. Watch your language! Racially loaded metaphors in scientific research. *BioScience* 57: 104–105.

Herbers, J. M., and S. Foitzik. 2002. The ecology of slavemaking ants and their hosts in north temperate forests. *Ecology* 83: 148–163.

Hölldobler, B., and E. O. Wilson. 1990. *The Ants.* Cambridge, MA: Belknap Press.

———. 1994. *Journey to the Ants.* Cambridge, MA: Belknap Press.

Hoyt, E., and T. Schultz, eds. 1999. *Insect Lives.* New York: John Wiley and Sons.

Johnson, C. A., and J. M. Herbers. 2006. Impact of parasite sympatry on the geographic mosaic of coevolution. *Ecology* 87: 382–394.

Maeterlinck, M. 1930. *The Life of the Ant.* New York: John Day Co.

Ratnieks, F.L.W., and P. K. Visscher. 1989. Worker policing in the honeybee. *Nature* 342: 796–797.

Ratnieks, F.L.W., and T. Wenseleers. 2005. Policing insect societies. *Science* 307: 54–56.

Sleigh, C. 2003. *Ant.* London: Reaktion Books.

———. 2007. *Six Legs Better.* Baltimore: Johns Hopkins University Press.

Smith, A. A., and K. L. Haight. 2008. Army ants as research and collection tools. *Journal of Insect Science* 8: 71–76.

Smith, A. A., B. Hölldober, and J. Liebig. 2009. Cuticular hydrocarbons reliably identify cheaters and allow enforcement of altruism in a social insect. *Current Biology* 19: 78–81.

Visscher, P. K., and R. Dukas. 1995. Honey bees recognize development of nestmates' ovaries. *Animal Behaviour* 49: 542–544.

Wenseleers, T., and F. L. W. Ratnieks. 2006. Enforced altruism in insect societies. *Nature* 444: 50.

Wheeler, W. M., and T. Schneirla. 1934. Raiding and other outstanding phenomena in the behavior of army ants. *Proceedings of the National Academy of Sciences USA* 20: 316–321.

9. Six-Legged Language

Aleksiev, A. S., B. Longdon, M. J. Christmas, A. B. Sendova-Franks, and N. R. Franks. 2008. Individual and collective choice: Parallel prospecting and mining in ants. *Naturwissenschaften* 95: 301–305.

Beekman, M., and J. B. Lew. 2008. Foraging in honeybees — When does it pay to dance? *Behavioral Ecology* 19: 255–262.

Beekman, M., R. S. Gloag, N. Even, W. Wattanachaiyingchareon, and B. P. Oldroyd. 2008. Dance precision of *Apis florea* — Clues to the evolution of the honeybee dance language? *Behavioral Ecology and Sociobiology* 62: 1259–1265.

Cerdá, X., E. Angulo, and R. Boulay. 2009. Individual and collective foraging decisions: A field study of worker recruitment in the gypsy ant *Aphaenogaster senilis*. *Behavioral Ecology and Sociobiology* 63: 551–562.

Conradt, L. 2008. Group decisions: How (not) to choose a restaurant with friends. *Current Biology* 18: R1139–1140.

Conradt, L., and C. List. 2009. Group decisions in humans and animals: A survey. *Philosophical Transactions of the Royal Society B* 364: 719–742.

Conradt, L., and T. J. Roper. 2005. Consensus decision making in animals. *Trends in Ecology and Evolution* 20: 449–456.

Couzin, I. D. 2008. Collective cognition in animal groups. *Trends in Cognitive Sciences* 13: 36–43.

Couzin, I. D., J. Krause, N. R. Franks, and S. A. Levin. 2006. Effective leadership and decision-making in animal groups on the move. *Nature* 433: 513–516.

Crist, E. 2004. Can an insect speak? The case of the honeybee dance language. *Social Studies of Science* 34: 7–43.

Detrain, C., and J.-L. Deneubourg. 2008. Collective decision-making and foraging patterns in ants and honeybees. *Advances in Insect Physiology* 35: 123–173.

Dussutour, A., S. C. Nicolis, E. Despland, and S. J. Simpson. 2008. Individual differences influence collective behaviour in social caterpillars. *Animal Behaviour* 76: 5–16.

Dussutour, A., S. J. Simpson, E. Despland, and N. Colasurdo. 2007. When the group denies individual nutritional wisdom. *Animal Behaviour* 74: 931–939.

Dyer, J. R. G., C. C. Ioannou, L. J. Morrell, D. P. Croft, I. D. Couzin, D. A. Waters, and J. Krause. 2008. Consensus decision making in human crowds. *Animal Behaviour* 75: 461–470.

Franks, N. R., A. Dornhaus, C. S. Best, and E. L. Jones. 2006. Decision making by small and large house-hunting ant colonies: One size fits all. *Animal Behaviour* 72: 611–616.

Franks, N. R., J. W. Hooper, A. Dornhaus, P. J. Aukett, A. L. Hayward, and S. M. Berghoff. 2007. Reconnaissance and latent learning in ants. *Proceedings of the Royal Society B* 274: 1505–1509.

Gorman, J. 2006. Mr. Speaker, I'd Like to Do the Waggle. *New York Times*, May 2.

Hauser, M. D., N. Chomsky, and W. T. Fitch. 2002. The faculty of language: What is it, who has it, and how did it evolve? *Science* 298: 1569–1579.

Lindauer, M. 1957. Communication in swarm-bees searching for a new home. *Nature* 179: 63–66.

Maeterlinck, M. 1901. *The Life of the Bee.* New York: Dodd, Mead and Co.

Marquis, D. 1987. *Archy and Mehitabel.* Reprint of 1927 edition. New York: Anchor.

Munz, T. 2005. The Bee Battles: Karl von Frisch, Adrian Wenner and the Honey Bee Dance Language Controversy. *Journal of the History of Biology* 38: 535–570.

Nieh, J. C., L. S. Barreto, F. A. L. Contrera, and V. L. Imperatriz-Fonseca. 2004. Olfactory eavesdropping by a competitively foraging stingless bee, *Trigona spinipes. Proceedings of the Royal Society of London B* 271: 1633–1640.

Passino, K. M., T. D. Seeley, and P. K. Visscher. 2008. Swarm cognition in honey bees. *Behavioral Ecology and Sociobiology* 62: 401–414.

Pinker, S., and R. Jackendoff. 2005. The faculty of language: What's special about it? *Cognition* 95: 201–236.

Planqué, R., A. Dornhous, N. R. Franks, T. Kovacs, and J. A. R. Marshall. 2006. Weighing waiting in collective decision-making. *Behavioral Ecology and Sociobiology* 61: 347–356.

Pollick, A. S., and F. B. M. de Waal. 2007. Ape gestures and language evolution. *Proceedings of the National Academy of Sciences USA* 104: 8184–8189.

Rittschof, C. C., and T. D. Seeley. 2008. The buzz-run: How honeybees signal "Time to go!" *Animal Behaviour* 75: 189–197.

Schultz, K. M., K. M. Passino, and T. D. Seeley. 2008. The mechanism of flight guidance in honeybee swarms: Subtle guides or streaker bees? *Journal of Experimental Biology* 211: 3287–3295.

Seeley, T. D., and P. K. Visscher. 2008. Sensory coding of nest-site value in honeybee swarms. *Journal of Experimental Biology* 211: 3691–3697.

Seeley, T. D., P. K. Visscher, and K. M. Passino. 2006. Group decision making in honey bee swarms. *American Scientist* 94: 220–229.

Sherman, G., and P. K. Visscher. 2002. Honeybee colonies achieve fitness through dancing. *Nature* 419: 920–922.

Skorupski, P., and L. Chittka. 2006. Animal cognition: An insect's sense of time? *Current Biology* 16: R851–R853.

Smith, E. M., and G. W. Otis. 2006. Resolution of a controversy: Functionality of the dance language of the honey bee, Part I. *American Bee Journal* 3: 242–246.

———. 2006. Resolution of a controversy: Functionality of the dance language of the honey bee, Part II. *American Bee Journal* 4: 335–340.

Su, S., F. Cai, A. Si, S. Zhang, J. Tautz, and S. Chen. 2008. East learns from West: Asiatic honeybees can understand dance language of European honeybees. *PLoS ONE* 3: E2365.

Visscher, P. K. 2007. Group decision making in nest-site selection among social insects. *Annual Review of Entomology* 52: 255–275.

Wenner, A. M. 2002. The elusive honey bee dance "language" hypothesis. *Journal of Insect Behavior* 15: 859–878.

Wray, A. 2005. The broadening scope of animal communication research. In *Language Origins: Perspectives on Evolution,* ed. M. Tallerman. Oxford: Oxford University Press.

Wray, M. K., B. A. Klein, H. R. Mattila, and T. D. Seeley. 2008. Honeybees do not reject dances for "implausible" locations: Reconsidering the evidence for cognitive maps in insects. *Animal Behaviour* 76: 261–269.

Yang, C., P. Belawat, E. Hafen, L. Y. Jan, and Y.-N. Jan. 2008. *Drosophila* egg-laying site selection as a system to study simple decision-making processes. *Science* 319: 1679–1683.

INDEX